# Anunnaki Homeworld

## Orbital History and 2046 AD Return of Planet Nibiru

by
Jason M. Breshears

THE BOOK TREE
San Diego, California

ISBN 978-1-58509-134-8

Cover art & design
Jason M. Breshears

Cover layout & colorization
Atulya Berube

Interior layout & editing
Paul Tice

Published by
The Book Tree
P.O. Box 16476
San Diego, CA 92176
www.thebooktree.com
We provide fascinating and educational products to help awaken the public to new ideas and information that would not be available otherwise.
Call 1 (800) 700-8733 for our FREE BOOK TREE CATALOG

This work is dedicated to
those of us within these
walls fated to suffer retroactive
laws and policies used by
our keepers to continue our
imprisonments, and to my mother
Doris who has taught me to see
past the present and have
faith in humanity.

# Table of Contents

# Acknowledgements

A special thanks to three individuals instrumental in the completion of this work. The author's father, Dan, has spent a small fortune providing the author with hundreds of contemporary and historic texts that are cited throughout this and the author's prior works, *Lost Scriptures of Giza* and *When the Sun Darkens*. Paul Tice of The Book Tree has provided valuable insight and made possible the publication of this work. We are further grateful to Jane Eichwald for her work in editing and manuscript preparation.

Special Note: Throughout the text the author references a number of his other works, some of which are yet to be published at the time this work was released. If by chance a desired book cannot be found, please contact The Book Tree for more information.

# Foreword

This work restores what our scholars have lost. The *truth* about our planet's past and future. This author's prior work, *When the Sun Darkens*, concerned the history and 2040 AD return of planet Phoenix, a fragmenting and uninhabited world having little bearing on the present thesis. The fact is herein demonstrated that there is indeed a *second* wandering planet that occasionally visits our inner solar system, a gigantic broken and presently *populated* world that is now fast approaching Earth and will arrive at 2046 AD. This planet is NIBIRU, and its orbital history completely forged the unfolding of human events.

This is the *Anunnaki Homeworld* and its return was a constant fear of the ancients. The most magnificent architectural wonders from the Old World, the Great Pyramid and Stonehenge, as well as the earth's most archaic dating systems all remain as mute witnesses of the presence and orbital chronology of this alien planet. Even the modernly misinterpreted Mayan Long-Count system, when *accurately* interpreted, as will be shown herein, was a sophisticated countdown to the exact date of the return of NIBIRU and its Anunnaki occupants in 2046 AD. The popularized year of 2012 AD was never the end of the Mayan system, and this will be conclusively proven in this book.

The pyramid was not a tomb. Stonehenge was not a temple. It is not mathematically possible for 2012 AD to be the end of the Mayan Long-Count calendar. These are the assumptions of men trained to think one-dimensionally, those blind to the silent atavistic patterns appearing mysteriously in our grain fields that beckon us to search deeper into the messages of universal geometry. We are being warned. These warnings concern 2046 AD.

Whatever preternatural forces are at work behind these amazing crop formations, it is abundantly clear that the exact same formula employed in interpreting the three-dimensional calendrical geometry of Stonehenge I and II is also the method for understanding the *calendrical* messages of crop patterns.

Reader beware. . . what the masses believe and what this thesis demonstrates cannot both be true.

# INTRODUCTION

This is the third book of Jason Breshears published by the Book Tree, but as the publisher it is the first time I have been compelled to write an Introduction to his work. Jason is an incarcerated individual. Generally speaking, it is an unspoken rule among publishers not to publish the works of those convicted of crimes. But Jason proved to be a rare exception.

First, he spent five years ordering books from our company while writing and studying diligently (although I was not aware of his writing at this time). When he finally approached us with a proposal, I knew him to be a serious researcher and good customer. Knowing he was imprisoned, I was still hesitant but decided to at least look at his work and consider it on those merits alone. He was not in jail for murder, which would have ruled him out, but made a terrible mistake while still a teenager. He is still paying for it today, and is at least trying to contribute something meaningful to the world in spite of his past.

Secondly, upon receiving his work it was clear that he was not writing about his life or the reasons he was there. He was not trying to create sympathy, or to make money from anything related to his conviction. He was, and is, deeply interested in uncovering the mysteries of mankind's past and of the earth we inhabit. And the research was solid.

The advantages Jason has is that, unlike millions who write books, he has fewer distractions and more time to write. His research adds a tremendous body of knowledge for those interested in these subjects, and not only has he supplied this knowledge, he has created much of the artwork in each of his books, including the covers. You would be hard pressed to find someone who could match his work ethic and level of research in this, or many other, areas of study.

The Anunnaki is a legendary race that appears in the oldest documents preserved by mankind. They are said to inhabit an outermost planet that orbits our sun in an extremely elliptical orbit. Each time this planet returns and gets near to the earth, it creates havoc and cataclysmic catastrophes. It has not appeared for quite some time, but many researchers claim it will be arriving in our vicinity again soon. The work of Zecharia Sitchin, author of the Earth Chronicles series of books, focuses on the Anunnaki and their previous visits to earth. I knew Mr. Sitchin well, having spent about three years as a personal videographer for him in ancient places around the world, and published a book for him in 1997 called *Of Heaven and Earth*. One question Sitchin was asked, above all others, was "When will this planet be returning?" He knew of the window in which it would return, but never gave an exact date. His concern was that people would dismiss all of his research should the planet fail to return during the exact year he might forecast, due to variations in such an elongated orbit – but he did say it was more than half-way back on its return journey.

Jason Breshears fearlessly provides an exact year, and does so confidently, due to his extensive research. Sitchin and others were never aware of this information, so it may add to the knowledge being sought in this area. Sitchin is not here to assess this work, but I believe he would acknowledge at least some of it as being valid and very important. And if you knew Sitchin, you know that is saying a lot.

The validity of this work will be determined by future events and, for now, by the opinions of you, the reader. It delves not only into historical records, as Sitchin had done, but also uses scientific cycles and mathematical formulas that relate to our conceptions of planetary time and orbits. The historical records, chronologically presented by Breshears, help in identifying cyclical patterns. He also employs advanced geometry, interprets complex crop circle patterns, and uses biblical prophecy in support of some of the astronomical events he has predicted. There is not a great deal shared on purely mythical aspects, like the Anunnaki "gods" themselves, but more on the actual sciences that can predict their return. Ancient stories and documented proof are two different things. For these reasons, I highly recommend this book.

Paul Tice

## *Archive 1*

## Existence and Return of NIBIRU

". . .the world of astronomers is in a state of terrorism,
though of a highly attenuated, modernized, devitalized
kind. Let an astronomer see something that is not of the conventional,
celestial sights, or something that is 'improper'
to see—his very dignity is in danger."

—Charles Fort, 1919, *Book of the Damned*

The world famous scholar Zechariah Sitchin began his campaign to inform humankind of the coming of the Anunnaki with the publication of his first work on the subject in 1976, entitled *The 12th Planet*. This and his following works clearly unveil a complex and rich history of powerful extraterrestrial beings that visited Earth through the agency of their roaming homeworld, called NIBIRU, a gigantic planet that according to Sitchin, nears our own planet every 3600 years. This scholar, able to translate the archaic Sumerian texts, has awakened millions to the possibility that Earth will be soon visited by these beings again. In this work this author will show that Zechariah Sitchin, a man with a name derived from that of a biblical prophet who foretold of the conditions in the Last Days, is himself a prophet whose ministry of divulging information on the Anunnaki actually began exactly *70 years* before these alien beings would return to Earth via their vagabond planet in 2046 AD. Much sooner than even Sitchin anticipated.

*Seven* years after *The 12th Planet* was released, NASA's Jet Propulsion Laboratory reported that their infrared telescope on the spacecraft IRAS had discovered beyond Pluto an ". . .extremely far mystery celestial body." It was said to be about the size of Neptune, or four times the size of Earth, and *moving in our direction*.(1) Six months later the planet was again detected. This intriguing discovery was made in the 207th year of the United States, this sum being half a Cursed Earth period, which is a system itself derived from the orbital longevity of another planet that Sitchin has never written about – that being planet *Phoenix*. In this author's prior work, *When the Sun Darkens: Orbital History and 2040 AD Return of Planet Phoenix*, is explained that there is a smaller planet that orbits the Sun every 138 years, called Phoenix by the ancients and represented in early Euphratean iconography as a *disk with wings*. This planet's entire history is independent from that of NIBIRU's and the two planets have often been confused with one another in the interpretation of traditions, myths, lore and ancient texts.

In the same year that these sightings occurred, they were also *silenced*. It is this author's contention that what was seen was in fact planet *Phoenix*, which had reached aphelion in 1971 AD, the furthest orbital distance from the Sun, way out in the Kuiper Belt region. When IRAS picked up the image, Phoenix was already in its 12th year back toward Earth on its highly elliptical orbit, to transit between Earth and the Sun in 2040 AD. Though Sitchin and a host of following authors claim this to be evidence of NIBIRU, the sighting is actually evidence of Phoenix. As we will see in this work, NIBIRU does *not* orbit along the plane of the ecliptic, nor anywhere near it.

9

In 1983 a think tank began of University of Chicago scientists who gathered evidence of past catastrophes, noting mass extinctions at regular intervals.(2)

The research concluded and the scientists submitted their findings in 1984 to two independent research think tanks of astrophysicists and astronomers who sought to analyze and interpret the data. The two separate groups arrived at the *same* conclusions. . . that our solar system either once was a binary system having two stars, or *still has a sister star*, a dark star known also as a compressed star. The scientists dubbed this dark star as Nemesis or the Death Star.(3)  An interesting tangent need be suffered here. In this same year of 1984, China and Britain signed an agreement returning Hong Kong back over to China, with China agreeing to maintain Hong Kong's capitalist system until the year 2047 AD.(4)  Do the British know something the Chinese don't? As we will see, after 2046 AD the world will be a very different place indeed.

Though scientists have concluded that Earth's chaotic past is the result of the presence and interaction of a sister star, there has been no official attempt to locate it. The existence of a Black Star, also called black holes, is not an untenable idea. These compressed stars are created when a normally luminous star's nuclear fuel has been expended (atomic fusion). With no more outward-directed pressure, the luminary darkens and begins folding in on itself with intense inward pressure of gravity resulting in an *implosion* of the star's mass, forming the beginning of a black hole.(5)  Now, this is a traditional physics approach, an abbreviated explanation serving to demonstrate how a star begins dying, however, the actual creation of a black hole may take a phenomenally long time to occur. The darkened star could remain for thousands of years or more as a black shadowy orb. Our own Sun may be very close to a Dark Star and we would not be able to see it, for telescopes and infrared instruments require that light enter through their lenses, but *no light can escape* a collapsing dark star due to immense gravitation.

Theoretical physics proposes a very radical yet plausible idea. Just as there exist particles and antiparticles, matter and the elusive dark matter, these subatomic particles (bosons, electrons, neutrinos, protons, positrons, gluons, neutrons, quarks, tachyons, gravitons and fermions) merely emulate their astronomical and colossal counterparts: luminous stars and dark stars, what were once called *frozen stars*. (6)  These darkened stars are spherical concentrations of mass to such a critical extreme that the space-time around them is warped so that nothing, including light photons, can escape it. These are of course the exact opposite of luminaries that give off light and fill the heavens. These compressed stars are invisible to astronomers because telescopes only see by virtue of escaping photons. Dark stars can come in all sizes. With no way to perceive their existence outside complex and often erroneous mathematical formulae that only take under consideration known variables, what we know of Dark Stars is still entirely theoretical.

But astronomers know that *something* is out there, for its unseen presence can be seen by the way it affects those things we can study. Everything is connected by gravity. All stars, planets, Moons, asteroids and comets are gravitationally connected to one another and to everything else. The Sun has a center of gravity along its equator known as the ecliptic plane, the imaginary line through the heavens whereupon the planets Mercury, Venus, Earth, Mars, the Asteroid Belt, Jupiter, Saturn, Uranus, Neptune and other objects like Pluto in the Kuiper Belt orbit the Sun. The travel through this plane would be more level in a perfect single-star system, but the truth of the matter is that many of our planets move under and above this ecliptic plane. Further, the orbits of the planets are *elliptical* and do not travel in true circular distances around the Sun, denoting that they are affected by an *outside* gravitational influence. The orbits of Mercury, Pluto, the asteroids and comets all have an extreme inclination from the ecliptic plane.(7)  The perturbations of the planets Uranus and Neptune absolutely require the presence of a nearby and powerful object gravitationally attracting the outer rim planets in our solar system.

Astronomers admit that more than half of the stars we see in the night skies through telescopes are *binary systems*.(8)  Incredibly, most binary systems are *differential binaries*, two or more stars revolving around one another of entirely different compositions.(9)  Now it must be recognized that any binary system will, by necessity, have two *ecliptics*, with each star maintaining their own center of gravity along their own equatorial mass. In fact, a vex unto modern scholarship is the recent discovery of a *triple*-system of stars, a trinary, with a gigantic *Jupiter-sized planet* orbiting the largest of the three stars in a system called HD188753 in the constellation Cygnus. This triad system is approximately 149 million light years away from Earth and the stars of this unusual system are *850 million miles* from another. From the surface of this huge planet can be seen one bright Sun and two far away smaller stars that shine as the most brilliant of stars.(10)  There is no evidence that multiple star systems maintain equal-plane ecliptics.

That another ecliptic plane intersects our own is further evident by the unusual and thus far unexplained 23-degree angle of the Earth's tilt. As the Earth travels around the Sun we would expect the planet to roll around its parent star with its axis pointed at a 23-degree inclination toward the Sun all year long, but this is not the case. For half of the year our planet's tilt is toward the Sun, while on our elliptical path the tilt points *away* from the Sun for half a year and still at *23 degrees*. Why is Earth's polar axis not locked in a 90-degree position as it orbits the Sun like Mercury and Venus?  Beginning with Earth and through the other planets of Mars, Jupiter, Saturn, Uranus and Neptune, all are tilted. The answer is glaringly evident. Mercury and Venus remain gravitationally locked to their parent star, but Earth, Mars and everything beyond the Asteroid Belt is powerfully influenced by the gravitational attraction of a nearby star.

The Dark Star's ecliptic plane intersects directly between the orbits of Venus and Earth virtually *perpendicular* to the plane of the sun's ecliptic upon which we orbit. This fact alone is why the ancient Sumerians called the Anunnaki Homeworld NIBIRU the *Planet of the Crossing*, as well as the designation *The Ferry*, because it was a planet that orbited *both stars*. The further away from the intersection of the binary ecliptics, the more evidence we see of the Dark Star's gravitational influence upon our own system. Earth's Moon literally rolls around our planet, its far side never seen and yet, as it travels, approximately every 28 days it suffers a drag that makes its own path irregular and even over time, inconsistent. A perfect Moon would eclipse the Sun, casting in shadow the exact same regions of earth time and time again with such predictive precision that eclipses would be published decades in advance as occurring only along a single strip of the earth's surface. But this does not occur, for the shadow of eclipses travels in winding curves and upon different regions of earth. We calculate them today because we take into consideration these unusual motions of the Earth and Moon, but the explanation of *why* we experience these motions is still forthcoming. Even the Moon's presence is a mystery, for it is a different composition than Earth, and in relation to the planet it orbits, our Moon is the largest in the entire solar system. Scientists cannot explain how such a colossal satellite could even be there without crashing into Earth.(11)

In this author's prior work, *When the Sun Darkens*, the entire orbital history of a planet known to our ancestors as *Phoenix* was provided, a planet actually seen by modern astronomers in 1764 AD when the planet transited between earth and the Sun and obscured one-fifth of the surface of the fiery orb and was seen by the naked eye. The astronomer Hoffman studied Phoenix and noted with astonishment that it traveled from the north to the south passing *over* the ecliptic plane and out of sight.(12)  This is exactly where NIBIRU transits and passes over the ecliptic, directly between Earth and Venus; however, NIBIRU is of a much more vast size than Phoenix. And unlike Phoenix, NIBIRU is *inhabited*. The inclined axis of Earth at 23-degrees, as well as those of the other planets, seem to remain aligned with the Dark Star's ecliptic plane. Additionally, unlike any other known Satellites in our solar system, two of Jupiter's Moons also orbit it from *pole to pole*.(13)

That a large, dark, opaque body was seen outside our system has been documented. . . and subsequently silenced. In the *Proceedings of the National Academy of Sciences 1915-394*, we learn that

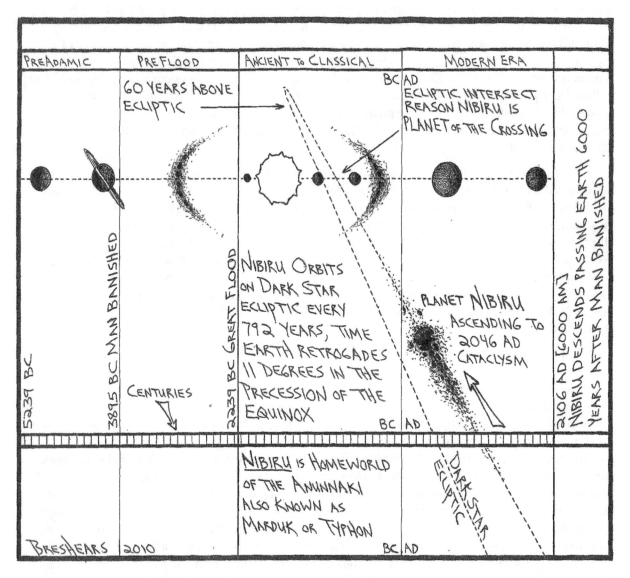

| PREADAMIC | PREFLOOD | ANCIENT TO CLASSICAL | MODERN ERA | |
|---|---|---|---|---|

60 YEARS ABOVE ECLIPTIC →

BC AD
ECLIPTIC INTERSECT REASON NIBIRU IS PLANET OF THE CROSSING

5239 BC

3895 BC MAN BANISHED

CENTURIES

2239 BC GREAT FLOOD

NIBIRU ORBITS ON DARK STAR ECLIPTIC EVERY 792 YEARS, TIME EARTH RETROGADES 11 DEGREES IN THE PRECESSION OF THE EQUINOX    BC AD

PLANET NIBIRU ASCENDING TO 2046 AD CATACLYSM

2106 AD [6000 AM]
NIBIRU DESCENDS PASSING EARTH 6000 YEARS AFTER MAN BANISHED

NIBIRU IS HOMEWORLD OF THE ANUNNAKI ALSO KNOWN AS MARDUK OR TYPHON

BRESHEARS 2010

DARK STAR ECLIPTIC

BC AD

a Professor Barnard discovered such a black body in Cephus. The very next year in the *Astrophysical Journal* of 1916 he modified his description to say as a "dark nebula."(14)  In 1919 Charles Fort, in his *Book of the Damned*, a critical analysis of scientific observations conveniently *forgotten* by the Establishment, he remarks that this dark companion of the star Algol was not the first time such blackened regions of space were seen and reported. Algol means *The Ghoul*, referring to a hellish creature whose fate was delayed; an *undead* being like the Anunnaki. Fort wrote . . .

> "Our acceptance is that vast celestial vagabonds have been excluded by astronomers. . . because they have not been seen so very often. The planets steadily reflect the light of the sun: upon this uniformity a system that we call Primary Astronomy has been built up; but now the subject matter of Advanced Astronomy is data of celestial phenomena that are sometimes light and sometimes *dark*, varying like some of the satellites of Jupiter, but with a wider range. However, light or dark, they have been seen and reported so often that the only important reason for their exclusion is – that they don't fit in."

Fort's work is filled with examples of phenomena in our own system that betray the influence of something else.(15) Andy Lloyd has recently published a work entitled *Dark Star* wherein he cites Dr. John Anderson, who believes that there is a planet with an orbital period of 700-1000 years that enters our own system perpendicular to the ecliptic.(16) This modern scientist is closer to the truth than either Sitchin, Lloyd or any known author today concerning the orbital longevity of planet NIBIRU. As will be revealed in this work, NIBIRU's orbit between both stars is exactly *792 years*.

Of epic importance is the 2005 discovery and first direct photograph of an *exoplanet*, the astronomical designation given to planets seen afar off that do not orbit any *known* star. This planet is twice the size of Jupiter and its distance in 2005 was reported at about 100 AUs away. An AU is an Astronomical Unit, about 93 million miles, the distance from Earth to the surface of the Sun. The most distant official planet in our system is Neptune, at about 30 AUs away from the Sun. As Earth travels around the Sun it eventually arrives to the opposite side of the luminary, 186 million miles away from where it started *six months* earlier, thus Earth travels around the Sun the distance of virtually *three AUs a year* in its orbit. Thus, 70 AUs is not far at all. In fact, the light of the Sun reaches Earth in 8 minutes, traveling at 186,000 miles per second. Past Neptune, sunlight would reach this giant planet 70 AUs away in 9 hours and 20 minutes, which isn't even 1% of a *light year*. The closest luminaries to our own solar system are Alpha Proxima, which revolves around its sister binary companion star, Alpha Centauri, at 4.2 *light years away* from our Sun. Jupiter is 318 times the mass of the Earth and is 5.2 AUs away from the Sun, orbiting it in just under 12 years. In this time this massive planet travels about 14 AUs in a solar system believed to span only about 30 AUs. One hundred AUs is nothing. The Alpha Proxima/Centauri system, being 4.2 light years away, is actually *252,000 AUs away*, for one light year equals 63,000 AUs. In *New Worlds in the Cosmos*, astrophysicists Michel Mayor and Pierre-Yves Frei reveal that gigantic Jupiter-sized planets have been discovered orbiting other suns in vastly irregular elliptical orbits. Exoplanet HD 80606 b is over four times the size of Jupiter, orbiting around the star Upsilon Andromeda with a periastron on 1.6 AUs with an orbit taking it out 3.4 AUs away from its Sun. NIBIRU's elliptical orbit is no longer theoretical, but demonstrative. What do these facts indicate? Again the evidence is clear. This huge planet is only called an *exoplanet* because the Dark Star it orbits has not been scientifically acknowledged, and 70 AUs is so close (much nearer than the Dark Star) that this recently discovered planet is within *our own system*. It could not possibly belong to other stars *light years* away.

NIBIRU will return in 2046 AD and to assume that a planetary body cannot travel 70 AUs in 41 years (from 2005 AD discovery) would be ridiculous, considering that Earth itself moves the distance of three AUs a year traveling around the Sun. During these 41 years Earth will have moved over *120 AUs* along the ecliptic. Not only is NIBIRU moving back toward Earth and the inner system since reaching aphelion in 1680 AD, but its own gravitational influence is being profoundly felt by our planet. In 1831 AD the North Magnetic Pole was first measured (151 years after NIBIRU began its return journey). In 2006 scientists asserted that the magnetic north pole has wandered some *500 miles* since it was first measured in 1831. The wandering was minute, initially, but in the last few decades the polar wandering has increased to *25 miles a day*.(17) As of 2008 NIBIRU is 38 years away. How much more will the magnetic wandering occur before something disastrous happens? Perhaps the unusual climatic changes transpiring around the world today are specifically the result of the approach of NIBIRU.

Sitchin's studies of the Sumerian records have persuaded him that NIBIRU is still inhabited by the Anunnaki, a pre-human race of virtually superhuman beings that visited earth in the distant past. These beings are actually described in the Book of Revelation and their contention against mankind is the background for much of the Old Testament and prophetic books. These Anunnaki are the Fallen Ones, a species of God's creation once loyal to their Maker but anciently banished, their Daystar *darkened* and, destined to only periodically visit our system, their planet suffering the force of entropy as it decays and fragments. Their planet is their prison and, as we will review, NIBIRU's own moon became a prison to the kings of the Anunnaki, a lost moon that has also been witnessed by the denizens of Earth.

The memory of the Anunnaki and their prison planetoid is the source for some profound mystical traditions and legends that have survived millennia to be recorded by pen and ink. Secretive cults, mystic orders and diligent antiquarians have for centuries preserved olden pieces of memories of this alien planet and its effects on Earth when the two worlds have nearly collided in passovers of catastrophic magnitude. Sitchin and all of those authors that followed him have done the subject justice and unearthed from the past hundreds of traditions, stories and ancient sightings that, to mention again, would only be redundant. The following evidence is only what Sitchin and the majority of researchers after him have neglected to mention.

Mystical records maintained even today assert that "radiant beings from heaven," came from a "dark red planet," a world of "sorrow and cataclysm."(18)  The Records of Thoth, according to modern occultists, claim that these beings are the ". . . spawn of a *lower star*.(19)  One of the most penetrating mystic traditions of this Anunnaki planet was recorded in 1880 AD by the occultist Thomas Burgoyne, whose collection of records and writings is within his *Light of Egypt* volumes.

Burgoyne wrote that there existed an unbroken tradition among mystics that there was a missing planet and that it was ". . .pushed out of line of march by disturbing forces," and that it ". . . became for a time, the prey of disruptive action and ultimately lost form, and is now a *mass of fragments*. The ring of planetoids between the orbits of Mars and Jupiter, indicates to us an empty throne."(20)  He also writes that this lost planet is the origin of what he terms as the *Dark Satellite*.

Burgoyne's mystical tradition seems to be confirmed by an Akkadian cylinder seal that depicts the solar system as having 11 planets. These would be Mercury, Venus, the Moon, Mars, the Lost Planet (Asteroid Belt), Jupiter, Saturn, Neptune, Uranus, Earth and NIBIRU. The Akkadian seal depicts a large planet between Mars and Jupiter where today lies an immense strewn ring of asteroids. The size of this strange planet on the seal is comparable to Jupiter and it dwarfs earth considerably.(21)  This Asteroid Belt is *not* NIBIRU, but the residual of *Phoenix*, which orbits every 138 years ever since it was shoved out of its orbital belt in 4309 BC (see *When the Sun Darkens*). Phoenix is heavily fractured and when it transits between Earth and the Sun it bombards our world with detritus, dusts and cometary fragments.

This information on the planets was evidently known in remote antiquity quite widely, for it is also represented upon a star chart of the Egyptian city of Thebes. This Theban chart shows all the known planets in our system, each symbolized with a bark (a boat representing its trajectory around the Sun like Jason on the Argo around the Zodiac). The planets of the inner solar system have barks because their orbits were known to the Egyptians, but the outer planets were not designated with barks because their motions were now known. The most intriguing aspect of this Theban knowledge was the fact that an additional bark of great size is depicted that conveys to us another planetary body with an extremely long orbital path.(22)  Some researchers like to believe that this chart identifies Pluto, but this is probably not the case. The planet seen in the Theban depiction may be of Phoenix. Pluto is three times smaller than our Moon and is so far away (varying between 2.8 to 4.6 *billion* miles as its orbits) that even the Hubble space telescope has only provided us blurred images. The Great Bark of this Theban astronomical chart would be NIBIRU.

It appears that the ancients would have us know that NIBIRU does not orbit the Sun as other planets do. It is highly elliptical and *not* on the sun's ecliptic. If it moved along the ecliptic, no matter how far, it would have been detected and recorded by modern astronomers. Very few stargazers pay attention to the stellar regions away from the plane of the ecliptic because it is here within this belt in the sky that the Sun and Moon traverse, as well as the planets Mercury, Venus, Mars, Jupiter and Saturn and the outer rim worlds. The Zodiac lies along this ecliptic belt. Probably the greatest deterrent to the discovery of this lost planet is due to its presence in the far southern heavens, which is not visible from anywhere north of the earth's equatorial regions. Even Zechariah Sitchin has written that NIBIRU would be seen from the southern heavens first and would *ascend* from below the solar system.(23)  The leaders of the scientific world for the past few centuries have been in Europe, Russia, the United States, Canada, China and Japan and *none* of these regions can view the southern heavens. In classical and more ancient times astronomers viewed the depths of space from the Mediterranean, North Africa, the Near East, India and China and still the southern circumpolar stars could not be seen. The Great Pyramid site is 30 degrees *north* of the Equator. Only from geographical areas south of the equator can the southern regions of space be viewed. This is the Deep, the *Abyss* of the Sumerians from which the Anunnaki emerged.

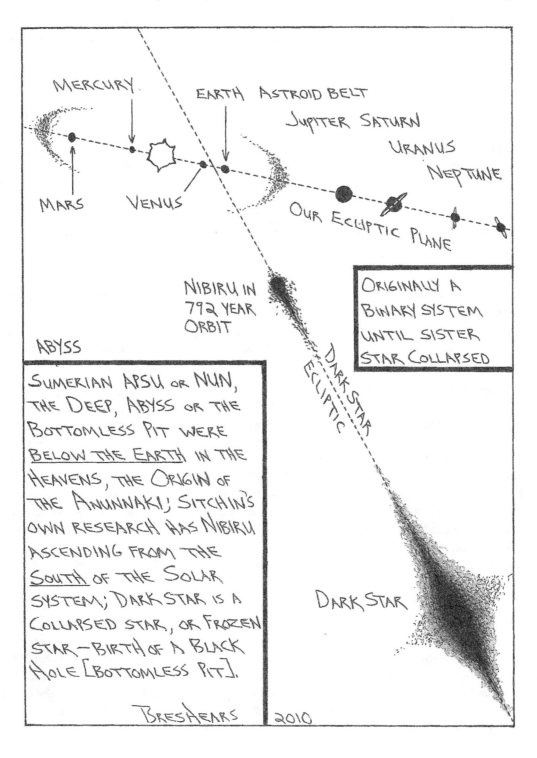

MERCURY.    EARTH    ASTROID BELT
JUPITER  SATURN
URANUS
NEPTUNE
MARS    VENUS    OUR ECLIPTIC PLANE

NIBIRU IN
792 YEAR
ORBIT

ABYSS

DARK STAR ECLIPTIC

ORIGINALLY A
BINARY SYSTEM
UNTIL SISTER
STAR COLLAPSED

SUMERIAN APSU OR NUN,
THE DEEP, ABYSS OR THE
BOTTOMLESS PIT WERE
BELOW THE EARTH IN THE
HEAVENS, THE ORIGIN OF
THE ANUNNAKI; SITCHIN'S
OWN RESEARCH HAS NIBIRU
ASCENDING FROM THE
SOUTH OF THE SOLAR
SYSTEM; DARK STAR IS A
COLLAPSED STAR, OR FROZEN
STAR — BIRTH OF A BLACK
HOLE [BOTTOMLESS PIT].

DARK STAR

BRESHEARS    2010

Burgoyne's writings are unique enough for they were published 96 years prior to Sitchin's. Burgoyne wrote that wicked beings still reside within this dead planet, as well as within its former moon, known as the Dark Satellite. He says that the Dark Satellite is inhabited by beings possessing the highest forms of cunning and intelligence ". . .who are the producers of the greatest portion of suffering and misery which afflicts humanity..."(24) Also, the rulers of the Dark Satellite ". . .mercilessly distort every arcane truth into theological dogma of partial error, causing it to assume in the human mind the delusive form of the externals of truth and logic. . . and corrupt Truth in every form wherein it struggles to become manifest on Earth.(25)

These records are stunning, for Burgoyne had no access to the Sumerian texts Sitchin studied, nor is there any evidence Burgoyne read Sumerian or ever heard the term *Anunnaki*. Hardly anything was known at this time about Sumer and the texts that had been excavated were still in museum basements awaiting translation. His information truly came from traditions and sources passed down outside the realm of modern scholarly media.

Another incredible parallel is mentioned inadvertently by Burgoyne. Sitchin wrote that the Anunnaki came to earth in need of gold to prolong their existence. Burgoyne wrote that the evil inhabitants of the Dark Satellite benefit by their wicked activities against humanity which provides them a ". . .means of prolonging their external existence upon the earth," for many of the former residents of the Dark Satellite have *descended to earth* already.(26) Burgoyne links the mystic traditions of the Dark Satellite to the Elder prophecies of the Vikings, known as the Ragnarok – Nordic predictions of the Last Days which will be ushered in by the return of an *unknown planet*.(27) Other Ragnarok fragments concern the return of Great Comets, (28) which will be seen in this book, that will accompany NIBIRU and even periodically break away from its surface to enter into their own solar orbits, as NIBIRU departs our own system.

This planet, destined to return to the inner system, was known by many names. It was called Tiamat and Marduk in the Near East, Typhon by the Greeks and Egyptians, Flying Serpent (Xiknalkan) by the Maya, (29) Antichthon by the Pythagoreans, (30) the Dark Star by the Hopi, who expect it to appear in the sky to initiate the end of the Fourth World, (31) the Dark Sun of other Native America groups and the Black Sun of the alchemists.(32)

After reviewing the mystical echoes of Burgoyne's research, perhaps one should be a little less critical of some similar modern satanic literature further substantiating the existence of the evil Anunnaki. In the *Testimony of the Mad Arab* as recorded in the occult book entitled *Necronomicon* (Tome of the Names of the Dead) we read, "May the gods be ever merciful to thee! May thou escape the jaws of the MASKIM, and vanquish the power of the Ancient Ones! And the gods grant thee death before the Ancient Ones *rule the earth once more!*"(33) The MASKIM is a Sumerian ideogram for a *race of demons* in olden tablet texts often linked to the Anunnaki.(34) The Ancient Ones are demonic Lords of the Anunnaki known in Sumerian and Babylonian texts and the Book of *Revelation* as the Seven Kings. They were present on earth and began a reign over mankind in 2839 BC before the Flood, their regnal period cut off by the Deluge, in its 600th year. The *Necronomicon* text appears to be derived from the Babylonian *Enuma Elish* tablets concerning the Anunnaki. Such a race of imprisoned beings is also the subject of the 5th century BC traditions of the Zervanites of Persia, who held that the star constellations were friendly to mankind, but the *planets* were actually *chained demons*.(35)

Continuing with the *Necronomicon* we read in the *Book of the Conjuration of the Watchers*: "And it is said that some of that race lie waiting for the Ancient Ones to once more rule the cosmos, that they may be given the right hand of honor, and that such as these are lawless. This is what is said."(36) The Watchers are the Anunnaki, this arcane title given to them in the Enochian works, the biblical book of Daniel and the Dead Sea records discovered in 1947 AD. They descended among men in the epoch before the Deluge and took the daughters of men to create a hybrid species known in the Scriptures as the Nephilim, the sons of the Watchers, a race of Giants.

As revealed in the writings left behind by Enoch and the Hebraic *Jubilees* text, the Watchers rebelled against the Creator over the issue of mankind and, in turn, God made men the weapon by which He would bring judgment against the Anunnaki. This is further reflected in the Necronomicon's *Magan Text*, reading that, "Man is the Key by which the Gate of IAKSAKKAK may be flung wide, by which the Ancient Ones seek their vengeance, upon the face of the Earth, upon the offspring of Marduk (Mankind). For what is new came from that which is old, and what is old shall replace the new. And once again the Ancient Ones shall rule upon the face of the Earth.(37)

Sumerian and Babylonian scholars are not unaware of these same traditions, for they fill up hundreds of archaic Euphratean texts. Our ancestors remembered the region of the Seven Kings of the Anunnaki and their 600-year reign over mankind before the Flood, a subject provided in depth in this author's work entitled *Descent of the Seven Kings: Anunnaki Chronology and 2052 AD Return of the Fallen Ones.*

If this thesis is to have merit, then there must be instances when in times of remote antiquity, as well as more contemporary times, the Anunnaki Homeworld had passed close to earth with its attendant satellites and comets, affected our own planet, and was recorded by men. There must be a pattern that remains fixed over the passage of millennia, over the distances of time and cultures to support the rich and global traditions of such a wandering planetary body. Critics beware, for these standards will be met herein abundantly.

Thousands of historical writings unveil nothing of NIBIRU's orbital longevity. Sitchin, Burgoyne nor any of the hosts of researchers and authors since them have provided an orbital chronology of the Anunnaki world. But unto the reader it is now rendered, through a lithic tapestry of tablet inscriptions, rock calendars like Stonehenge and the chronoliths of the Great Pyramid. Among the most startling proofs of the return of NIBIRU in 2046 AD is found within the amazing mathematical framework of the *Mayan Long-Count calendar*.

Unto you is opened a door into the near future. Sitchin lit for us a torch, and now we carry it through this threshold of the present to study what will come to be from the records of the past.

# Archive 2

## Orbital History of Planet NIBIRU

". . .what a distance there is between a truth that is glimpsed and a truth that is demonstrated."

—Quoted from French admirer of Isaac
Newton in *The Discoverers*, by Daniel Boorstin

In order to demonstrate any object's orbital history one must first retain an accurate chronology of the world's history. This author has provided this in his *Chronicon: Timelines of the Ancient Future*, an expansive chronological work detailing our planet's history and *future* from 5239 BC to 2106 AD that aligns and correlates over *forty* different ancient and modern calendars and dating systems. The exact year-dates in the timelines of *Chronicon* are further supported by the precise and meticulously-charted year-dates encoded within the Great Pyramid's geometrical calendar as seen in this author's work entitled *Chronotecture: Lost Science of Prophetic Engineering*. Additionally, verifying the exactitude of this timeline is the orbital chronology of planet Phoenix as shown in *When the Sun Darkens: Orbital History and 2040 AD Return of Planet Phoenix*. Within these works are provided the hundreds of sources that validate these timelines.

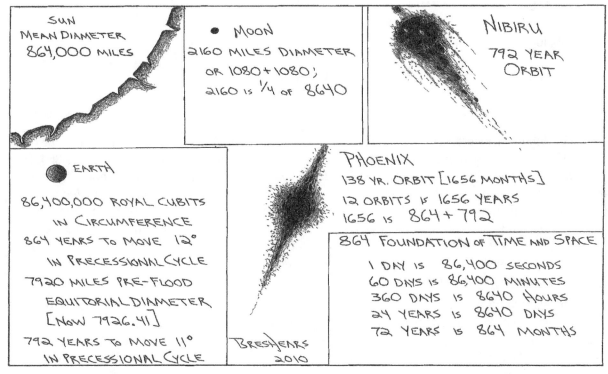

SUN
MEAN DIAMETER
864,000 MILES

• MOON
2160 MILES DIAMETER
OR 1080 + 1080;
2160 IS ¼ OF 8640

NIBIRU
792 YEAR
ORBIT

● EARTH
86,400,000 ROYAL CUBITS
IN CIRCUMFERENCE
864 YEARS TO MOVE 12°
IN PRECESSIONAL CYCLE
7920 MILES PRE-FLOOD
EQUITORIAL DIAMETER
[NOW 7926.41]
792 YEARS TO MOVE 11°
IN PRECESSIONAL CYCLE

PHOENIX
138 YR. ORBIT [1656 MONTHS]
12 ORBITS IS 1656 YEARS
1656 IS 864 + 792

864 FOUNDATION OF TIME AND SPACE
1 DAY IS 86,400 SECONDS
60 DAYS IS 86,400 MINUTES
360 DAYS IS 8640 HOURS
24 YEARS IS 8640 DAYS
72 YEARS IS 864 MONTHS

BRESHEARS
2010

19

In this author's prior work, *When the Sun Darkens*, the Annus Mundi Chronology is explained in detail, an unbroken timeline from *year one* to the 6000th year of Man's Banishment from Eden. Year One is thoroughly demonstrated to be 3895 BC and the Great Flood that virtually destroyed the earth occurred 1656 years later in 2239 BC (1656 AM). The 6000th year when the Anunnaki are to be vanquished by the Chief Cornerstone (Stone the Builders-Anunnaki Rejected) is 2106 AD, known as Armageddon. It is a universal belief that the Watchers/Nephilim/Fallen Sons of God/Anunnaki caused the Flood during a time when civilization was centered in Egypt, a culture of prediluvian people having no connection to the dynastic Egyptian civilization celebrated today which emerged *after* the cataclysm. It happened that 792 years after this Deluge, the tribes of Israel in Egypt were led out of the country by Moses after a series of cataclysms in 1447 BC (2448 AM), known as the Exodus.

That a 792 year period counted down forward to this monumental and biblical event is recorded in the chronometry of the Great Pyramid, a geometrical calendar that demonstrates over and over that dates recorded within its own timeline are reflected both forward and backward to similar occurrences, not only in history but also in *architectural features*. And this proves interesting to our present study. With the Flood as 1656 Hebrew Reckoning and 1656 Annus Mundi, or 792 + 864 years (1656), we are provided the key to Sumerian sexagesimal calendrical mathematics, as well as an insight to the connection between Earth and *NIBIRU*. The number 864 is considered by many numerologists to be the *foundation of time* number, which links all the world's calendrical systems that are predicated on units of 6, 12 and 144. The sum of 864 is itself 144 x 6 or 108 x 8, both 144 and 108 being prominent in the chronometry of the Great Pyramid and *Stonehenge*, as will be reviewed. The diameter of the Sun in miles is 864,000 and this Earth that travels around it has an equatorial diameter of *7920 miles* (792 x 10). However, the circumference of the Earth in royal cubits used to measure out the Great Pyramid is exactly 86,400,000 cubits.(1)

The sum of 864 is the common denominator between time and *space*:

| | |
|---|---|
| 864 | months in 72 years (72 is half of 144) |
| 8640 | days in 24 years (360-day years) |
| 8640 | hours in 360 days (ancient year) |
| 86,400 | minutes in 60 days |
| 86,400 | seconds in 1 day (1 day is 1440 minutes) |

In Hebrew the words *world* and *all nations*, as well as *habitation*, all have a gematrical value of 432.(2) Combining these terms provides us with the phrase *whole inhabited world* for a value of 864. In Stonehenge the Anunnaki are represented in the presence of the *bluestone* ring and horseshoe. The ring has a diameter of 79.2 feet.(3) Conceptually we can interpret this data to show that the *whole inhabited world* (space beginning with a time of 864) would be destroyed (Flood) by addition of Anunnaki (duration of 792 years) ending in 1656th year (864+792), or 2239 BC. Our planet and that of the Anunnaki are related in *space and time* more than any other planets in this solar system. In fact, it will be shown in this book that NIBIRU as well as Earth *do not belong in this system*, both having their origin with the ancient star, now dying, once called the Daystar, but now has gone dark.

In this author's work entitled *Descent of the Seven Kings: 2052 AD Return of the Fallen Ones*, we are shown the Anunnaki Chronology timeline, a calendar beginning with Anunnaki history in 5239 BC that divides our own world's history into 600-year long epochs. In the 600th year of the Anunnaki Chronology, or 4639 BC, the Anunnaki rebelled against their Maker over the issue of Mankind. In 4309 BC the Daystar imploded, becoming a Dark Star, and billions of rock, ice and planetary detritus passed through the present solar system and damaged all the planets and moons, even destroying the *original* Phoenix (leaving behind a ringed Asteroid Belt). The dead and fragmenting shell of Phoenix now orbits on the Dark star's ecliptic, orbiting our Sun every 138 years. Our own planet traveled 270 years from its orbital position around the Daystar to its *present* position – tightly fit between the orbits of Mars and Venus, where we do *not* belong. In the equidistant-exponential distribution of the planets from the surface of the Sun, the presence of our planet between Mars and Venus disrupts the entire harmonic pattern, but this is the subject of *Descent of the Seven Kings*. In 4039 BC mankind was placed in the Garden of God ("walled enclosure" in Hebrew) and the Builders (Anunnaki) set to the task of renovating Earth. In the 6th day, according to Genesis 1, was man created, or 144 hours (24 x 6). And in the 144th year after he was made, mankind rebelled against the word of God and was cursed in 3895 BC (Year One, Annus Mundi). In *Descent of the Seven Kings* it is conclusively shown that in the Anunnaki's 1800th year (600 x 3) they *descended* on Earth among mankind, with this year being 3439 BC (456 AM).

The 3439 BC descent of the Watchers is the primary focus of the *Book of Enoch*, a text that not only details their arrival, their mission, sexual integration and creation of genetic hybrids, but also their *fate* to be imprisoned and then released during the Last Days to execute these exact same designs *again*. The *Book of Enoch* even opens up in its first passage with a notice to the reader that the text was specifically recorded for the time of the end. As the descent of the Anunnaki in 3439 BC was 1200 years (600 + 600) before the Flood in 2239 BC, this period in *days* was exactly 432,000 (half of 864,000). Before the cataclysm of 713 BC, when the Sun retrograded 10 degrees during a near-collision with the Dark Satellite, as we will review, Earth orbited the Sun in 360 days. After 713 BC the orbit was lengthened by 5.25 days a year, resulting in an annual period of 365.25 days. This 1200-year period from their descent to the Deluge is featured in many major ancient legends and myths of the Ages prior to a dreadful cataclysm, with these stories related in *Descent of the Seven Kings*. For our purposes here we revert back to our erudite scholar Zechariah Sitchin, who wrote that the Sumerian tablet texts he translated specifically record that the Anunnaki and the planet NIBIRU arrived and passed Earth exactly 432,000 years before the Flood.(4) As unveiled in *Descent*, scholars have misinterpreted the years and *shars* of Sumerian reckoning as annual periods of the Earth around the Sun, but the truth of the matter is that the Sumerians referred to planetary movement as shars, as Genesis 1 reveals, ". . .the evening and the morning was the first *day*." The 432,000 shars is thus 432,000 days, or precisely *1200 years*. Inadvertently, Sitchin too, provides us with the date of 1200 years before the Flood, 3439 BC. This is not unprecedented, for the Mayan Long-Count and many others like the Kali Yuga all factored *days*, and any association to years is through misinterpretation. The year was insignificant to the pre-flood people because prior to the catastrophe there were no *seasons*. Even the early prophetic texts like *Enoch* wrote of the Last *Days*, not years.

This year is confirmed in another way. According to the *Book of Jubilees* (4:15), the father of Enoch was born four years after the appearance of the Watchers on earth. He was named Jared in commemoration of their arrival; his name derived from the root 'ared, meaning *descent*. Jared was born in 460 Annus Mundi, or 3435 BC, four years *after* 3439 BC. This amazing year appears to be the primary astronomical focus of the Great Pyramid's *descending* passage. This tunnel leads into the Subterranean Chamber area below the edifice to a pit and, as detailed in *Chronotecture* and *When the Sun Darkens*, these architectural features refer to the Abyss, the Bottomless Pit below the solar system from where NIBIRU emerges every 792 years. The antediluvian pole star was Alpha Draconis, the Eye of the Dragon, with the earth enjoying a 90-degree axis like Mercury and Venus and revolving around the Sun in a perfect 360-day year, aligned in synchronicity with the Zodiacal Year. In the 1870s Astronomer Royal of Scotland, Charles Piazzi Smyth, concluded in his study of the Great Pyramid that the Descending Passage, for reasons unknown to himself, had pointed directly at *Alpha Draconis* in the year *3440 BC*, (5) this being only one year off the true date of 3439 BC. Alpha Draconis was the circumpolar constellation, the Lord of the Houses of the Zodiac. In the precessional cycle, the entire zodiacal belt retrogrades 1 degree every 72 years, and 11 degrees is *792 years*, the Anunnaki in the Babylonian *Enuma Elish* tablets being represented by the number 11. Also, 12 degrees of the Zodiac requires *864 years*.

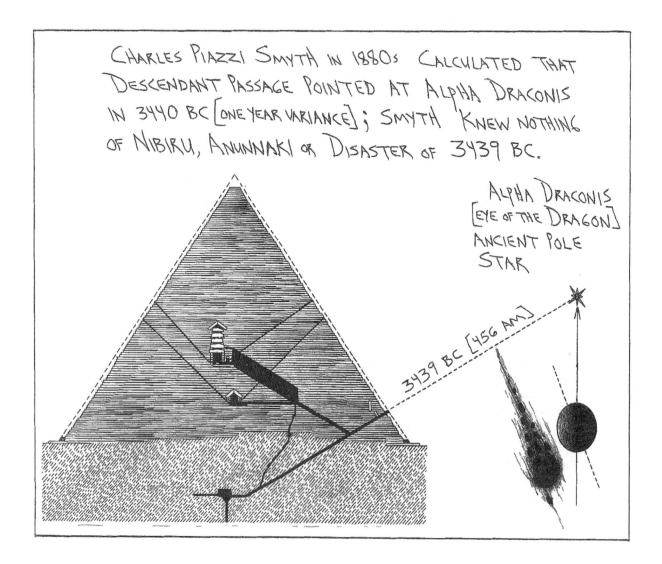

CHARLES PIAZZI SMYTH IN 1880s CALCULATED THAT DESCENDANT PASSAGE POINTED AT ALPHA DRACONIS IN 3440 BC [ONE YEAR VARIANCE]; SMYTH KNEW NOTHING OF NIBIRU, ANUNNAKI OR DISASTER OF 3439 BC.

ALPHA DRACONIS [EYE OF THE DRAGON] ANCIENT POLE STAR

3439 BC [456 AM]

The beginning and end of the passovers of NIBIRU are interlinked and involve the Anunnaki. The history of the world, rise and fall of civilizations, comet impacts, meteoric fallout, plague mists, earthquakes and mass migrations are the direct result of the continual presence of NIBIRU in our system and the debris the Anunnaki Homeworld leaves in its wake.

## Orbital History of Planet NIBIRU

3439 BC
(456 AM)

As NIBIRU descends over ecliptic, 200 Watchers arrive on Earth by a mutual compact to create a race of genetic hybrids to help them establish their control over mankind. They conspire to provide humanity with advanced knowledge in exchange for their *daughters*. Females are provided to the Anunnaki and in return, mankind is given specific knowledge of herbalism, sorcery, incantations, power over evil spirits (ghosts of Pre-Adamic people), metallurgy, weaponsmithing, how to make war machines, jewelry, feminine cosmetics and brew draughts that would render women barren (contraceptives) so as to retain their beautiful figures. They were also taught advanced astronomical sciences and astrology and divination.(6)  The Anunnaki instructed men in the Building Craft. In this year the world suffered terrible earthquakes and flooding. The Jewish Haggadoth writings tell of this flood which was more ancient than Noah's (in 2239 BC), which occurred in the reign of Enosh.(7)  The Yezidis of Asia also recall this belief in two archaic floods, specifying that Noah's was the later one.(8)  As this is the exact timing of NIBIRU according to the Sumerian texts, and Enosh was indeed king according to the *Book of Jasher* in this year, the following excerpt proves useful to our thesis:

> "The sons of men forsook the Lord all the days of Enosh and his children; and the anger of the Lord was kindled on account of their works and abominations which they did in the earth. And the Lord caused the waters of the Gihon river to overwhelm them, and he destroyed and consumed them, and He destroyed the *third part of the earth*, and not withstanding this, the sons of men did not turn from their evil ways. . . .(9)

The Gihon is the Nile river in Egypt and the reader must remember here that the passing of NIBIRU not only brought the Anunnaki to Earth, but also simultaneously resulted in the death of a *third* of mankind. This is the underlying theme of the return of NIBIRU in the Last Days. The Anunnaki Homeworld entered the system in 3499 BC (396 AM), unseen from Earth, and required 60 years to pass over the system until it descended in 3439 BC to bring the Anunnaki as it passed close to Earth, finishing its 792 year orbit between *both stars*. Thus, NIBIRU's orbital period is below our system's 732 years (366 to aphelion and back) and 60 years *over* the ecliptic for a total of 792 years. This pattern remains *fixed* throughout history. Sitchin remarks in *Wars of Gods and Men* that the Sumerian texts over and again claim that the starting point for their histories was exactly 432,000 "years" before the Great Deluge. As these are actually *days*, this pinpoints the year as 3439 BC, or 1200 years before the Flood in 2239 BC. This is the year, according to the *Jasher* account, that a flood destroyed a third of the world, and again Sitchin's studies confirm this. Sitchin cites a text excavated from the ruins of Sumer concerning the descent of a god named EN.KI that reads, "When I approached Earth, there was much *flooding*." In *Hamlet's Mill*, Giorgio de Santillana and Hertha von Dechend note that the number 432,000 seems to be where science and myth merge, so that this number is significant in mythological systems around the world, as noted duly by Sitchin in citing their work.

| | |
|---|---|
| 3073 BC (822 AM) | NIBIRU reached aphelion (furthest point from the Sun) after 366 years, starting 366 years back toward inner system. 200th year of Enoch's life. |
| 2707 BC (1188 AM) | NIBIRU ascends out of the Deep after 732 years (366 + 366), entering inner system unseen from Earth. This is 108 Anno Pyramid, which started with Great Pyramid's completion in 2815 BC (1080 AM). NIBIRU begins 60 years over solar ecliptic. |
| 2647 BC (1248 AM) | NIBIRU descends over ecliptic, completing *792-year* orbit (732 + 60) unseen from Earth. |
| 2281 BC (1614 AM) | NIBIRU reaches aphelion after 366 years, starting 366 years back toward inner system. |

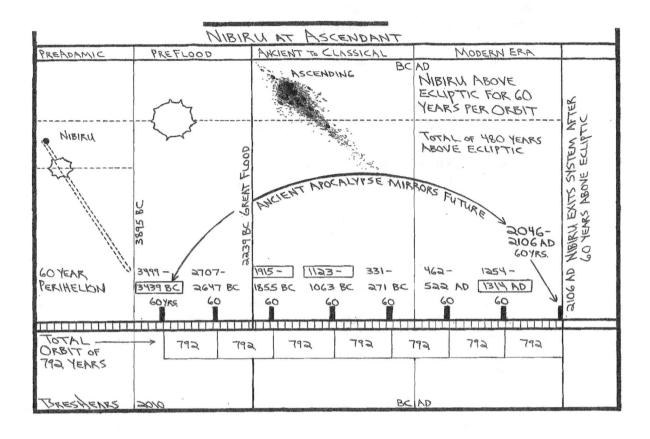

NIBIRU AT ASCENDANT

| PREADAMIC | PREFLOOD | ANCIENT TO CLASSICAL | MODERN ERA | |
|---|---|---|---|---|

ASCENDING

BC | AD

NIBIRU ABOVE ECLIPTIC FOR 60 YEARS PER ORBIT

TOTAL OF 480 YEARS ABOVE ECLIPTIC

ANCIENT APOCALYPSE MIRRORS FUTURE

2046 – 2106 AD 60 YRS.

2106 AD NIBIRU EXITS SYSTEM AFTER 60 YEARS ABOVE ECLIPTIC

NIBIRU

3895 BC

2239 BC GREAT FLOOD

60 YEAR PERIHELION

3499 – 3439 BC 60 YRS.

2707 – 2647 BC 60

1915 – 1855 BC 60

1123 – 1063 BC 60

331 – 271 BC 60

462 – 522 AD 60

1254 – 1314 AD 60

TOTAL ORBIT OF 792 YEARS

| 792 | 792 | 792 | 792 | 792 | 792 | 792 |
|---|---|---|---|---|---|---|

BRESHEARS 2010

BC | AD

---

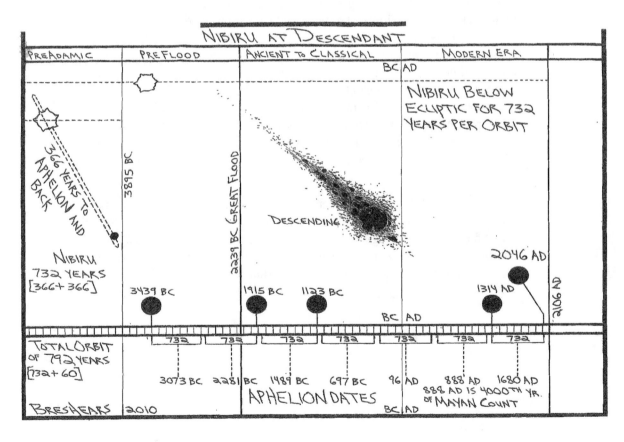

NIBIRU AT DESCENDANT

| PREADAMIC | PREFLOOD | ANCIENT TO CLASSICAL | MODERN ERA | |
|---|---|---|---|---|

BC | AD

NIBIRU BELOW ECLIPTIC FOR 732 YEARS PER ORBIT

366 YEARS TO APHELION AND BACK

NIBIRU 732 YEARS [366 + 366]

3895 BC

2239 BC GREAT FLOOD

DESCENDING

2046 AD

2106 AD

3439 BC

1915 BC

1123 BC

1314 AD

BC | AD

TOTAL ORBIT OF 792 YEARS [732 + 60]

| 732 | 732 | 732 | 732 | 732 | 732 | 732 |
|---|---|---|---|---|---|---|

3073 BC   2281 BC   1489 BC   697 BC   96 AD   888 AD   1680 AD

APHELION DATES

888 AD IS 4000TH YR. OF MAYAN COUNT

BRESHEARS 2010

BC | AD

1915 BC
(1980 AM)

NIBIRU ascends out of the Deep after 732 years (366 + 366), entering the inner system. Sargon I of Akkad reigned, later to be known as AMAR.UDA.AK (Marduk) of Sumer and Babylon, called Amraphel and Nimrod by the ancient Hebrews. The Anunnaki planet appears close to Earth as a great black shadow *darkening the stars* and emerging from the *underworld* (below solar system). The Babylonians see the black planet as a celestial dragon spitting fire and lightning upon Earth called Tiamat, a huge aberration in the night sky that trembles the Earth that the Egyptians referred to as Typhon. Because it passes and everything returns to normal after the atmospheric and geologic disturbances, Sargon I was considered a vanquisher of the *Anunnaki*, and Marduk in Babylon was famed for returning a constellation back into the sky – NIBIRU merely moved away and the constellation reappeared. These traditions are even immortalized in the *Enuma Elish* epic. Hesiod's *Theogony* further preserves this memory, saying that it occurred shortly after Zeus (Nimrod) conquered the Titans (Japhethites: see *Chronicon* at 1948 and 1920 BC). Hesiod wrote that Typhon appeared with strange *dragon heads* (olden designation for *comets*), with *black* tails. The dragonheads gave off flames and vibrated the earth with noises as mountains trembled. Thunderbolts struck the earth and water boiled until a fiery hurricane was born. Earthquakes shook the world and the sea overextended its bounds. The disaster (this word means *evil star*) concluded with a powerful lightning bolt flashing between earth and Typhon, which gave rise to the innumerable legends of a Hero God who slew a Sky-Monster with his Labrys (Thunderbolt) or Battleaxe.(10) This "lightning bolt" is none other than a flux tube of electricity between the two planets, a phenomenon first witnessed by astronomers in modern times in 1994 AD when comet Shoemaker-Levy 9 fragmented into about 21 pieces and slammed into Jupiter. The close proximity of a metal and rock moon of NIBIRU to Earth causes this object to break free of NIBIRU's gravitational hold. This begins the orbit of the *Dark Satellite*, which will pass through the inner system for 16 years until nearly colliding with Earth again in 1899 BC before moving to a much more stable orbit. The Dark Satellite will be covered later in this work. This appearance of NIBIRU is consistent with the period when Hammurabi (Amraphel-Nimrod) initiated the Code of Laws famously named after him, which venerate the Anunnaki in their opening lines. This year is 900 Anno Pyramid and 324 After Flood (108 x 3).

1855 BC
2040 AM

NIBIRU descends over ecliptic, completing 792-year orbit (732 + 60) unseen from Earth. A massive strewn field of rock and comet debris trailing the dead planet breaks free and begins its own orbit around the Sun as the *Sodom-Trojan Apocalypse Group*. This begins a 2520 (360 x 7) day countdown to the destruction of Sodom & Gomorrah and the Harappan civilization of the Indus Valley.

1489 BC
(2406 AM)

NIBIRU reaches aphelion after 366 years, starting 366 years back toward inner system.

1123 BC
(2772 AM)

NIBIRU ascends out of the Deep after 732 years (366 + 366) in direct transit between Earth and the Sun, causing the Sun to *darken*. Due to ecological disasters and a dynastic change in China, Wu Chang establishes the Zhou Dynasty and 250,000 Chinese take ships and sail to *America*. This account is little known to readers in the West and interestingly, this year is 324 years (108 x 3) after Exodus (1447 BC Israelites depart Egypt).

1063 BC
(2832 AM)

NIBIRU descends over ecliptic completing 792-year orbit (732 + 60) unseen from earth.

697 BC
(3198 AM)

NIBIRU reaches aphelion after 366 years, starting 366 years back toward inner system.

331 BC
(3564 AM)

NIBIRU ascends out of the Deep after 732 years (366 + 366), entering inner system unseen from Earth in same year that Alexander the Great's Grecian force defeated the Persian army of Darius in northern Mesopotamia in the *Battle of Gaugamela*. Alexander also took Egypt and this start of a Greek Empire was the 600th year of the Divided Kingdom Chronology, counting the years from the division between Israel and Judah in 931 BC, and the 414th year (Cursed Earth period) of the Post-Exilic Chronology from 745 BC when the Ten Tribes of Israel were deported from Palestine and taken into Assyrian domains to later migrate into Persia, Asia Minor and Europe.

271 BC
(3624 AM)

NIBIRU descends over ecliptic completing 792-year orbit (732 + 60) unseen from Earth. An immense strewn train of debris trailing the Anunnaki Homeworld is captured in the Sun's own gravitational field, breaking free of NIBIRU, and begins its own orbit as the *Romanid Apocalypse Comet Group*.

96 AD
(3990 AM)

NIBIRU reaches aphelion after 366 years, starting 366-year journey back toward the inner system. While the imprisoned Anunnaki are at this furthest distance from the Earth, the Creator gives mankind the gift of the future knowledge concerning the return of the Anunnaki and their Seven Kings in the Last Days in the form of the vision known as the *Revelation*, recorded by the last apostle, John, a prisoner on the isle of Patmos. This man was essentially suffering on his small island what the Anunnaki suffered on their prison planet: confinement.

462 AD
(4356 AM)

NIBIRU ascends out of the Deep after 732 years (366 + 366), entering inner system unseen from Earth in the 2700th year (900 x 3) after the Great Flood.

522 AD
(4416 AM)

This is the ONLY year in history that NIBIRU *and* planet Phoenix traverse the inner system at the *same* time, both descending over the ecliptic. NIBIRU crosses ecliptic after 60 years orbiting over the Sun, and Phoenix ends its 138-year orbit to begin another one. See *When the Sun Darkens*. The immense train of NIBIRU yields forth a strewn field mass that begins orbiting the Sun and breaks into two distinct groups: the *2046 AD NIBIRU Comet Orbit Group* and the *Reuben Comet Group 2047 AD*. This year is 5760 Anunnaki Chronology, or 144 x 40 or 360 x 16.

888 AD
(4782 AM)

NIBIRU reaches aphelion after 366 years, starting 366-year journey back toward the inner system. Much of its train begins falling behind the planet.

1254 AD
5148 AM)

NIBIRU ascends out of the Deep after 732 years (366 + 366), entering inner system unseen from earth. Free from its gravitational hold and lagging behind by six years is an expansive detritus field that had, at some prior time, broken off of the dead planet's surface and will enter the inner system and disperse into several different comet groups in 1260 AD.

1314 AD
(5208 AM)

NIBIRU descends over ecliptic, entering the inner solar system in direct *transit* between Earth and the Sun in the 16th year of the historic period known to scholars as the Seven Comets of Europe. NIBIRU appears in the final year of a period when meteorites, comets, plagues and earthquakes from 1298-1314 AD decimated the populations of Asia, the Middle East and Europe (probably Africa, Americas too, but no records passed down). In the *Black Plague* by George Deaux we find that ". . .the first reports (of the plague) came out of the East. They were confused, exaggerated, frightening, as reports from that quarter of the world so often are: descriptions of storms and earthquakes; of meteors and comets trailing noxious gases that killed trees and destroyed the fertility of the land. . ." (11)  William Bramley in *Gods of Eden* wrote that when the comets passed, there were reports of plague mists from China to Europe and many historians estimate that a *third of Europe* was killed by the Black Plague.(12)  This, by extension, infers that the death toll was probably a *third of the entire world*, for many nations left records of these catastrophic events but actual death tolls were not taken in China, India and elsewhere outside of Europe. One of the sicknesses that developed was influenza, which derives from the word *influence* because it was believed to have had its origin with the *stars* (influence of the stars), or comets. NIBIRU transited and all of China and Asia and that hemisphere during the daytime was cast into utter *darkness* as the world trembled and quaked (13), but on the other side of the world at night, European historians recorded that an ". . .awe-inspiring *blackness*" appeared and passed over the heavens.(14)  This was the last appearance of NIBIRU in the inner system before it returns in 2046 AD, this historical period mirroring both the 3439 BC appearance of NIBIRU that destroyed a *third of humanity* and initiated flooding and disasters before the Deluge of Noah, and again in 1915 BC when NIBIRU appeared as a great black *darkness* in the heavens with *comets* and fiery storms which the ancients called Tiamat and Typhon. As NIBIRU has trailing it hundreds of millions of fragments of rock and ice, so too do we see that it also has a vanguard preceding it of comets and detritus in a strewn train. The plague fogs, disrupted weather patterns and blighting of crops is all remembered by the Greeks in an amazing tradition concerning the Telchines, a race who lived before the Flood and were guilty of carving the first images of the gods. Zeus decided to destroy them because they had been interfering with the *weather*, raining *magical mists* and *blighting crops*.(15)  The Telchines were enchanters worshipped by a pre-flood matriarchal society, these being the Anunnaki and their Genitrix (see *Descent of the Seven Kings*). Interestingly, the Greek myths of Typhon claim he *rose up* from under heaven to challenge the gods (other planets on ecliptic), and indeed in 1915 BC the Anunnaki Homeworld *ascended* into the inner system. This year is precisely *792 years* to the year 6000 Annus Mundi (2106 AD), Armageddon War when the Chief Cornerstone will vanquish the Anunnaki. This year of 1314 AD was 6552 (6000 + 552: Phoenix Cycle) of the Anunnaki Chronology, and 3552 (3000 + 552 years) after the Flood. In the following year, 1315 AD, bad harvests are reported throughout Europe.(16) An estimated 10% of the population died as cannibalism and starvation was rampant.

1680 AD
(5574 AM)

NIBIRU reaches aphelion after 366 years, starting 366-year journey back toward inner solar system.

2046 AD
(5940 AM)

NIBIRU ascends out of the Deep after 732 years (366 + 366), entering the inner solar system and nearly *colliding* into Earth. The effect NIBIRU will have on our world in this year of 2046 AD is so catastrophic, and so detailed in the archives of ancient prophetic literature, that all of this will await for a full description later in this book.

2106 AD
(6000 AM)

NIBIRU descends back over ecliptic close to Earth in the 6000th year of the Annus Mundi (original Hebraic) timeline from Man's Banishment in 3895 BC.

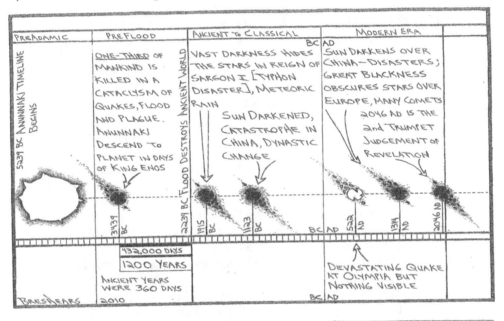

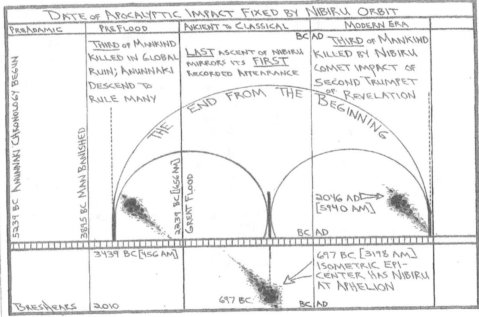

In order to fully appreciate the effect NIBIRU will have upon our world in 2046 AD we must first review how this Anunnaki Homeworld has affected our planet in the past through its distinct and devastating comet groups. As we will now find, the Anunnaki planet has been fragmenting and continually seeding our solar system with the mechanics of apocalypse.

## *Archive 3*

## Ancient Earth-Killer Comet Group

Comets are signs of the revolutions of kingdoms, the rebellions of cities,
famines, pestilences, tides of the ocean and of earthquakes. . .
—Stobaeus in *Physica*


Each time NIBIRU reaches perihelion in the inner solar system its deep, frozen-fractured oceans fragment and significantly sized regions break away from its surface. After traveling with the disintegrating planet through billions of miles of space, this world approaches perihelion near Earth and the areas of this growing strewn field of rock and cometary detritus are freed from NIBIRU's gravitational hold and enter into their own orbits around the Sun. Some of these groups maintain orbits mathematically connected to the orbit of NIBIRU, while others become more erratic. *All* of the groups experience entropy, begin disintegrating, fragment further, die out and spread out into wider and longer debris fields and trains.

The first group detailed herein broke away from NIBIRU in 3499 BC. This was exactly 60 years prior to the descent of the Anunnaki in 3439 BC, before the Flood, as the homeworld of these Watcher beings entered the system. In the 18th year of their passing over the Sun, planet Phoenix descends from over the Sun to south of the ecliptic in 3481 BC (see *When the Sun Darkens*), perturbing the comets, drawing them into a 138-year orbit around the Sun. This begins the amazing orbital chronology of the *Ancient Earth-Killer Comet Group* in 3361 BC.

| | |
|---|---|
| 3361 BC<br>(534 AM) | Comet group remains concentrated in a debris field all within one year of each other as they descend over the ecliptic, free of both NIBIRU and now Phoenix, stabilizing on a 396-year solar orbit linking these comets to NIBIRU, for 396 is exactly half of NIBIRU's 792 year orbit (396 x 2). This seemingly arbitrary date is directly indicated within the Great Pyramid's chronometrical scheme, for the entire length of the Ascending Passage to the Great Step (emblematic of Christ) is 3361 Pyramid Inches (see *Chronotecture*). |

(330) years below ecliptic)

| | |
|---|---|
| 3031 BC<br>(864 AM) | The comets ascend over the ecliptic in this amazing year of 864 Annus Mundi, the year representing the *Foundation of Time* of the Sumerian sexagesimal system. It is to be recalled that this sum's import relates to the movement of the Earth and is 792 years before the Great Flood. |

(66 years above ecliptic)

31

2965 BC
(930 AM)
The patriarch of humankind made in the image of God, *Adam*, died at age 930 of the Hebraic/Annus Mundi system (Genesis 5:5), however, his age is measured from the Banishment of Man from Eden in 3895 BC. Adam was created 144 years prior by the Builders (Anunnaki) at the behest of God when the Creator breathes His Spirit in him (which offended the Anunnaki) in the 6th day of the Renovation of Earth (144th hour) in 4039 BC. Just prior to his death, on his deathbed, Adam informs the Sethites of the coming of the Great Flood that will destroy the entire world. When Adam died the *Sun darkened* as a comet, for the first time in human history, transited and occulted the Sun. (1) The 930 years of Adam mirror the 930 years (5239-4309 BC) of the Pre-Adamic World governed by the Anunnaki until it was destroyed.

(330 years under ecliptic)

2635 BC
(1260 AM)
No records; 1260 is half of 2520 (360 x 7).

(66 years above ecliptic)

2569 BC
(1326 AM)
No records; this is the 40th year of the reign of the Anunnaki King Enmengalanna, the midpoint of his 80 year rule. He is one of the Seven Kings listed on the ancient Sumerian King Lists.

(330 years below ecliptic)

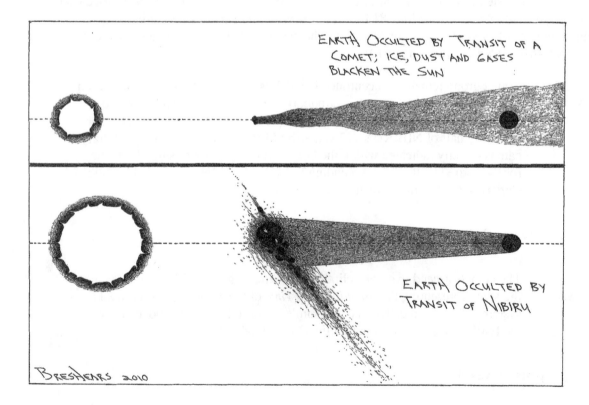

2239 BC
(1656 AM)

Planet Phoenix transits, *darkening the Sun* seven days before the Flood begins (see *When the Sun Darkens*). This year completed four Cursed Earth periods of 414 years each and three Phoenix Cycles of 552 years each (1656). As Phoenix transits and a large comet approaches Earth of the *Ancient Earth-Killer Comet Group*, earth suffers a pole shift according to the *Book of Enoch* (64:1-4) and the *Tractate Sanhedrin* at *108b*. The Rabbinical records indicate the Flood was caused by the falling of *two stars*.(2)  The number 108 archaically identified the *comet* as clearly revealed in this author's other works, *Descent of the Seven Kings* and *Chronicon*, which interestingly is referred to in *Tractate* 108b as well as on page 108 of Frank Joseph's book *The Destruction of Atlantis*, wherein he notes that it is believed that the Flood was caused by a comet. The antediluvian world enjoyed a pristine climate globally due to a watery atmosphere called the Firmament Above, a marine canopy that magnified the heavens while also reflecting away ultraviolet radiation from the Sun. The comet plunged directly into and *through* this mesosphere of water high in the sky and initiated its collapse, a virtual implosion of water that some believe was frozen smooth like glass, allowing the ancients to see incredibly vast distances in space. Earthquakes and strange electrical storms rage as the heavens are turned upside down and rains increase upon earth. Meteorites from the comet group are drawn in and bombard the surface, often trailing noxious gases before impacting in other regions and many die by asphyxiation. The foundations of the world (basement rock) are broken and pillars of lava ignite gases with boiling water that spews forth in thousands of fountains and geysers with volcanic violence not experienced by mankind since this date. A global ocean is created and the former surface materials of the planet are scrambled and much is reconstituted according to weight and density, with billions of dead people, Giants, and creatures from land and sea all get deposited into these layers as the immense pressure from the weight of this churning ocean and the boiling waters full of radioactive contamination entombs them into what is essentially *cooked* under these conditions into the layers of the fossiliferous and terribly misinterpreted geologic column. The Old World has been turned into a single *planetary fossil* of a former global ecosystem that had not known arctic polar regions or seasonal changes, with that earlier world having been a global greenhouse. The Flood transpired 792 years after 864 Annus Mundi (3031 BC), and began a 792-year countdown to the Exodus and Ten Plagues on Egypt in 1447 BC. This was the 3000th year of the Anunnaki Chronology, which began in 5239 BC (see *Descent of the Seven Kings*).

(66 years over ecliptic)

2173 BC
(1722 AM)

No records; Chinese tradition of Emperor Chang holds that the Sun darkened in what is translated as 2173 BC, this being 36 years after the Flood. Perhaps an anachronism.

(330 years below the ecliptic)

1843 BC          No records.
(2042 AM)

        (66 years over ecliptic)


1777 BC          No records.
(2118 AM)

        (330 years under ecliptic)


1447 BC          This is exact date of the Exodus of Israel from Egypt and the Ten Plagues.
(2448 AM)        Plagues break out all over Egypt (possibly the world) and Tacitus wrote that
                 the horrible epidemic disfigured the body.(3)  The Chinese *Mawangdui Silk
                 Almanac* tells of a Great Comet with Ten Tails (Ten Plagues origin?) that
                 Graham Phillips believes appeared in about 1500 BC.(4)  This was actually
                 in 1447 BC, only 53 years variance. Phillips further demonstrates that this is
                 what the Egyptians saw in the sky, ten tails that literally filled the heavens.
                 (5)  The comet's tail occults the Sun and in *2 Esdras* we read that the WHOLE
                 WORLD suffered an earthquake as heavenly omens ". . .troubled the men of
                 that age."(6)  The Jewish *Haggadoth* text reads that the Tenth Plague upon
                 Egypt was by far the worst, for the majority of the Egyptians died, expiring
                 in this plague.(7)  The incredible Egyptian Ipuwer papyrus reads that these
                 plagues and darkness was foretold to the ancestors.(8)  Ipuwer wrote that the
                 Sun was covered in darkness, people cowered in fear, blood was everywhere
                 and whole cities were destroyed.(9)  Earthquakes, meteoritic rain and noxious
                 storms were all attended by erratic behavior from animals and insect swarms.
                 A reddish contamination of the Nile and the violent volcanic eruption of Thera
                 from the Mediterranean also occurred. *Again*, Crete and Mycenaea are ruined,
                 this being 240 years (60 x 4) after they were previously destroyed in 1687
                 BC by the transit of Phoenix (see *When the Sun Darkens*) and giant skeletons
                 with colossal skulls have been excavated from among the ruins of Thera.(10)
                 Linear A Script of Minoan Crete ceases and the survivors develop Linear B by
                 adjusting other proto-Greek scripts. Many of the megalithic cities around the
                 world that were rebuilt after the 1687 BC cataclysm are now abandoned for
                 good, to be overtaken by the sea, by jungle growth, by desert sands, mined for
                 new building materials or naturally covered by the earth over the millennia.
                 This is the year 2448 Annus Mundi, the Hebrew Year 2448 being recorded
                 in the Talmudic *Seder Haddaroth* text as being the year of the Exodus and
                 Ten Plagues on Egypt.(11)  Interestingly, Patrick Geryl and Gino Ratincx
                 independently discovered that the ancient Egyptian number for *cataclysm* was
                 *2448*.(12)  The amazing synchronicity of this year and actual chronological
                 proofs that this was 1447 BC are to be seen abundantly in *Chronicon* (see 1447
                 BC).


        (66 years over ecliptic)

1381 BC
(2514 AM)
No records; this is the *108th year* of the life of Joshua, the Hebrew leader who commanded the Israelite army in the Conquest of Canaan in 1407 BC. In this year Joshua relinquished his command over Israel and passed over the rulership of the people to the Elders.(13)

(330 years under ecliptic)

1051 BC
(2844 AM)
By this time the comets are spread out into a train of three or four years. However, the 1063 BC entrance of NIBIRU in the inner system perturbs these comets and the immense strewn train is divided into two groups under the influence of tidal forces from NIBIRU, the Sun and a close pass by Earth. The larger group becomes the *King of Israel Great Orbit* comet group beginning in 1051 BC, this being King Saul of Israel's first regnal year, a wicked king. The comet group is spread out into a 14-year long train to 1039 BC, the year of David's birth, who will become King David, a living prophecy of the coming of Christ. The smaller detritus field will continue on, unperturbed on its established orbit.

(66 years over ecliptic)

985 BC
(2910 AM)
No records.

(330 years below ecliptic)

655 BC
(3240 AM)
No records.

(66 years over ecliptic)

589 BC
(3306 AM)
No records.

(330 years below ecliptic)

259 BC
(3636 AM)
The deteriorating strewn train of the *Ancient Earth-Killer Comet Group* enters the inner system in a 19-year long chain of debris.

240 BC
(3655 AM)
A comet is seen by the Chinese, (14) this being in the final year of the passing of this comet group. It will be seen a final time in 174 BC.

(66 years over ecliptic)

183 BC
(3702 AM)
No records; in this year the 19-year long debris train passes into the system, the 19th year being 174 BC.

174 BC
(3721 AM)

Paralleling the 19th year of the train in 240 BC, in this year during the Roman consulships of Spurius Postumius and Quitius Mucius, *two suns* appeared at midday according to Pliny.(15)  This describes what appears to be a disintegrating comet with light shining through and off of its millions of fragments. This is the end of the *Ancient Earth-Killer Comet Group.*

The next comet group explored is actually very much a part of this first one, for it was initially joined to it when these immense glacial sheets fragmented off the surface of NIBIRU and fractured into comets and asteroid chains. It wasn't until 1051 BC that the two groups separated, by a large margin due to gravitational strain and chaotic dynamics through the interaction of the Sun, NIBIRU's presence in the inner system and a close pass by earth.

## Archive 4

# King of Israel Great Orbit

"Everything occurs in accordance with Law, and *chance* is but Law unrecognized." —Kybalion

This comet group breaks away from the train of the *Ancient Earth-Killer Comet Group*, interestingly fulfilling a circular destiny, for this King of Israel Great Orbit, which broke free of NIBIRU long before the Flood as part of the train of the Earth-Killer group, will end in 2046 AD in conjunction with the *return of NIBIRU*. This scenario fulfills the concept that an end refers to its beginning. There is a definitive theme accompanying this comet group – that the *counterfeit* of Christ comes *first* in the unfolding of history.

| | |
|---|---|
| 1051 BC (2844 AM) | As the group passes unseen from Earth, the tribes of Israel in violation of ancestral decrees, elect a *king*. The first ruler of Israel is King Saul, who would fulfill the role of the Wicked King and be cursed (see *When the Sun Darkens*). This train of debris is 14 years long, stretching to the year 1039 BC, the exact year of David's birth, a boy who will slay the giant Goliath and later assume the kingship of Israel, establishing the Throne of David. NOTE: David is born in the 12th year (144 months) of the train, which is 14 years long to 1037 BC. Saul was chosen by the people and became corrupted, David was chosen by God. |

(66 years over ecliptic, still traveling with Ancient Earth-Killer Group before breaking away)

| | |
|---|---|
| 985 BC (2910 AM) | The 14-year long strewn train enters the inner system. |
| 973 BC (2922 AM) | Is 12th year (144 months) of the train. David, after Saul's death, fulfilled role of Wicked King by having a lover's husband killed in battle. Israel loses 70,000 people to a plague and this provides David atonement for the trespass and he is forgiven.(1)  Though the biblical records do not specify, the fact of the matter is that the plague was probably *global*. |

37

971 BC
(2924 AM)

In the 14th year of the debris chain King David *dies*.(2) His son Solomon becomes king of Israel. He will be the third and *last* king that Israel ever had before Israel and Judah become separate kingdoms in 931 BC.

This group that detached from the Ancient Earth-Killer Comet Group to become the King of Israel Great Orbit is drawn away into outer solar system and loses velocity, becoming what astronomers term a long-period comet (group). For the longevity of this group's orbit it will remain fixed at 984 years under the ecliptic and 526 years above the ecliptic.

(984 years below ecliptic)

1 BC
(3894 AM)

The Persian Magi having interpreted correctly the nine planetary-stellar conjuncts that transpired in the heavens in 3-2 BC (see *Chronicon* for specifics), now followed a strange moving *star* in the heavens, which was a comet of the *King of Israel Great Orbit* group which directs them to Bethlehem, where they find the infant Jesus (Yeshua). According to the Qumran records of the Dead Sea Scrolls found in 1947, a comet appeared at this time, in conjunction with a planetary conjunct in Pisces. Some believe the comet appeared in 4 BC but this is because of the popular myth of Christ's birth in 4 BC, which is without merit. The planetary conjunctions heralding the birth of the Savior occurred in 3-2 BC with the comet appearing in 1 BC. In the *Epistle of Ignatius to the Ephesians* we read, "How then was our Savior manifested to the world? A star shown in the heavens beyond all other stars, and its light was inexplicable, and its novelty struck terror into the minds of men. . . and men began to be troubled to think whence this star came so unlike to all the others." The Christ is not yet one year old. His *2 BC birth* is confirmed by the fact that:

a.   Tertullian wrote that Jesus was born 28 years after Queen Cleopatra died in 30 BC (30-28 is 2 BC);

b.   Irenaeus wrote that he was born in the 41st year of the reign of Augustus Caesar (Octavian), who came to power in 43 BC (43-41 is 2 BC);

c.   Vedic scholars assert that the Kali Yuga (Black Age) counted down 3100 years to the coming of a Savior. As these scholars claim the Kali Yuga calendar began in 3102 BC, 3100 years later is 2 BC. The December 25th date is NOT the birth of Christ, but the arrival of the Persian Magi who followed these prophecies and others like them. Jesus was born 120 days earlier.

The 14-year-long detritus train still holds true and is verified in the annals of history. In 526 AD Dionysius Exiguus calculated that Jesus was born in the 754th Year of Rome. He miscalculated by two years and further designed the Year of Our Lord Calendar (Anno Domini) to begin at AD 1, which was actually the 753rd year of the Roman Calendar, the city founded in 753 BC. Dionysius invented this system 526 years after it actually began; however, his mistake was divinely inspired. The BC-AD confluence is best understood when paralleled with the unbroken Annus Mundi Chronology:

4     BC.....3891 AM
3     BC.....3892 AM
2     BC.....3893 AM
1     BC.....3894 AM
1     AD.....3895 AM
2     AD.....3896 AM
3     AD.....3897 AM

14 AD
(3908 AM)

This is the 14th year of the comet train. At this time the Roman military on the German frontier began a full mutiny over their treatment and living conditions and was on the verge of broiling into violence when the *full Moon darkened* without a cloud in the sky according to Tacitus in the *Annals of Imperial Rome*. The Moon reappeared only to fade away again.(3)  The phenomenon distracted the disgruntled soldiers and a peaceful resolution was later adopted. In this same year during a gladiatorial show put on by Germanicus, a *meteor* sped across the sky in broad daylight, alarming the people.(4)  Caesar Augustus died (ruling when Christ was born) and was succeeded by Tiberius Caesar (ruling when Christ was Crucified).

(526 years over ecliptic)

526 AD
(4420 AM)

That the comet group should return in this year signifies an underlying scheme operative in the Plan of the Creator. This was the exact year Dionysius Exiguus designed the Anno Domini Calendar. On May 29th a massive earthquake rocked the Jewel of the East, the famed city of Antioch in Syria. As the ground trembled *flames fell from the sky* and a horrible stench bubbled up from the sea (hydrogen sulfide?) according to John of Ephesus.(5)  This cosmopolitan city caught fire and thousands of people trapped in the rubble were burned alive waiting to be dug out. In the days following the catastrophe organized bands of robbers and bandits plundered, raped and dispatched many of the survivors. Total death toll was estimated at 250,000 in Antioch alone, but the devastation included surrounding regions. The train maintains its 14 year-long pattern.

536 AD
(4430 AM)

A famine begins in England and spreads throughout Ireland, lasting until 538 AD. Plague spreads through Africa, Egypt, Syria and the east. This is the 19th year of the passing of the King of Israel Great Orbit.

538 AD
(4432 AM)

Emperor Justinian, a thoroughly wicked ruler over the Roman world whose life is detailed in Procopius's *The Secret History*, in this year declared the Pope in Rome to be the *Sovereign Ruler* over Christendom. According to the *Anglo-Saxon Chronicle* for this year a strange total eclipse of the Sun occurred and the stars appeared in the daytime.(6) Because astronomers know that there was no total eclipse to be seen in this year, the event has been relegated as fictional; however, this was no doubt the transit of a gigantic comet that occulted the Sun. The start of Papal rule over European kingdoms is thematically related to *oppressive kingship over Israel*, and the descent of the modern nations from the Lost Ten Tribes of Israel is clearly manifest in the annals and evidence is exhibited profusely in *Chronicon: Timelines of the Ancient Future*.

540 AD
(4434 AM)

*The Anglo-Saxon Chronicle* records that another total eclipse occurred. (7) And again, historians ignore this event simply because they assume that only the Moon occults the Sun.

542 AD
(4436 AM)

Although this is two years beyond the 14-year train, the plague of 536 AD had now spread throughout India and the Far East, having decimated Palestine, Syria, Asia Minor and now Byzantine. Procopius recorded that it killed thousands within the ancient city.

(984 years under the ecliptic)

1510 AD
(5404 AM)

The Aztecs recorded strange omens in the sky and a comet in the night sky. The lake around Mexico City began *boiling* and native sages interpreted these signs to mean that evil approached them. And 19 years later, Hernan Cortez and the Spanish conquistadors enslaved them.

(526 years over the ecliptic)

2036 AD
(5930 AM)

For an exhaustive treatise revealing how 2036 AD is the first year of the *Antichrists' reign* and the first year of the 70-year *Apocalypse* to 2106 AD (6000 AM), the reader is directed to review 2036 AD in *Chronicon* and in this author's other work entitled *Chronotecture: Lost Science of Prophetic Engineering*, the Great Pyramid's timeline. This is the FIRST SEAL of the Revelation. The Antichrist will seal Israel's approval by instigating a war with the nations of Islam by taking back by peace settlement all the geographical regions formerly under the dominion of King Saul, David and Solomon. In this year the Antichrist is 40 years old, born in 1996 AD when the comet Hyakutake passed over the star *Al Gol* (The Ghoul: Undead Being like Antichrist), a star called by the ancient Hebrews as the *Head of Satan*, a sign in heaven reconfirmed in 1997 on the *exact same day* (April 11th) when the comet Hale-Bopp also passed over the star Al Gol.(8)  Also confirming the 1996 AD birth of the Antichrist is a series of amazing crop circle formations in 1996-1997 AD (see *Chronicon*). Muslim nations will come together in one massive army of millions that will in two years invade Israel in what the Antichrist and his agents have specifically designed as a COUNTERFEIT ARMAGEDDON to occur in 2038 AD. The entire scenario is planned to deceive the peoples of the earth into believing that this person, the False Messiah, is in fact the returned *Jesus the Christ*. The apocalypse is brought upon man because of his *pride*.(9)  A comet of the King of Israel's Great Orbit will appear and be construed as a sign of *approval*, the group reentering the inner system since their last pass in 1510 AD when they served as a warning to the Aztecs of the coming of a False Savior (they believed Hernan Cortez was Quetzalcoatl: Mexican Savior, discovering their error only too late). The appearance of this same group in 538 AD proves interesting because Procopius wrote that he believed Emperor Justinian (who elevated the Pope to universal ruler over Christendom) was actually the *King of the Demons* in disguise. The 14-year train will pass until 2050 AD.

2038 AD
(5932 AM)

This begins the Second Seal of Revelation as hundreds of *millions* of Muslim soldiers in a jihad invade Israel to exterminate them. This is prophetically known as the invasion of Gog and Magog, foretold in Ezekiel 38 and 39 (38 paralleling this 20*38* AD date). When Ezekiel wrote this prophecy these nations listed in this passage had *nothing* in common, of various ethnic groups and religions, some matriarchal or patriarchal with different gods and pantheons and some were even enemies. But EVERY nation mentioned in this invasion passage against Israel today is *Islamic*. As the countless horde advances upon Israel, a comet of the King of Israel Great Orbit enters the atmosphere and explodes over the masses of men, literally vaporizing them – similar to the comet detonation in 1908 AD over Tunguska. The event will be construed as a miraculous deliverance and the Antichrist who initiated the break in the Peace Covenant and caused the War will be given credit for the destruction of the Muslim armies. The prophet-priest Ezra foresaw this when he wrote about the armies of Arabia advancing upon Babylon in the Last Days (Babylon will be seat of Antichrist's power) as well as toward the west (Israel), when the appearance of a *horrible star* (comet) emboldens them before they are burnt up in the fire.(10)

| | |
|---|---|
| 2040 AD (5934 AM) | In this, the 4th year of the comet train, planet Phoenix, after 138 years orbiting the Sun since 1902 AD, returns in direct transit occulting the Sun and Moon, causing a pole shift as meteorites bombard Earth, fulfilling the *Sixth Seal* (see *When the Sun Darkens*). |
| 2042 AD (5936 AM) | Cometary detritus rains on Earth. (see *Chronicon*). |
| 2044 AD (5938 AM) | Meteorites bombard the earth and set entire region on fire in the *First Trumpet* judgment. This fallout could be from both the King of Israel Great Orbit and the strewn fields preceding the arrival of NIBIRU in 2046 AD, two years later. (see *Chronicon*). |

While the affect of this comet group may be seen on Earth for a few more years to 2050 AD, we end our review of this time period until later, after we have had exhausted the orbital chronologies of other relevant comet groups seeded by NIBIRU. This group is still passing Earth in 2046 AD when the Anunnaki Homeworld passes and nearly collides into our planet, these comets having their ultimate origin as fragments of seas and oceans once upon NIBIRU's surface.

## Archive 5

# Orbital History of the Dark Satellite

"According to the unanimous opinion of scholars, the circumference of the whole earth, which seems to us immense, is no more than a tiny point in comparison with the vastness of the universe."

—Ammianus Marcellinus, *Book 15* (4th cent. AD)

The Dark Satellite is a lost moon of NIBIRU serving the Creator as an orbiting *prison* containing those ancient creations that rebelled against Him. Not all of them, for the core masses of the Anunnaki remain entrapped upon NIBIRU, but as Enochian records indicate, a select prison was made for those leaders among the Watchers that bound themselves by mutual compacts and curses when they descended in times before the Flood to begin their oppression and deception of mankind. The infamous Seven Kings of the Anunnaki are imprisoned with these denizens within this dark moon. NIBIRU entered the system in 1915 BC and trailing the Anunnaki Homeworld was its lost moon, which by consequence of its lagging distance, now initiates its own orbit around the Sun and nearly collided into Earth in 1899 BC. It will maintain an orbit mathematically related to the 792-year orbit of NIBIRU, being 395.4 years initially, but due to the near collision of 713 BC the orbital velocity was altered to 395 years, this being virtually half of 792 years (790).

1899 BC
(1996 AM)

This year is the foundation to worldwide traditions and memories concerning the erection of an immense Tower that reached into the heavens and was destroyed. After a few decades of construction the Tower in Shinar (called the Tower of Babel) was almost complete. This massive brick pyramid was built at the order of Nimrod (Amraphel/Hammurabi) to commemorate the Flood and stand in defiance of God. The Tower was in fact a giant pyramid in Mesopotamia. As the project drew to a close, the *Dark Satellite* crossed the ecliptic in a near collision course with Earth. Earthquakes shook the world, the masonry was damaged to the extent that it could not support the upper weight and internal collapse occurred, with thousands of people still working in and out of the structure while at the same time a powerful thunderbolt, called a flux tube, a blast of lightning discharged between the Dark Satellite and Earth, vitrified a third of the falling masonry while similar lightning blasts of immense energy vitrified entire areas from Asia Minor, the Middle East, northern Egypt and India. Forts, cities, smaller towers, great walls and whole areas by virtue of the thunderbolts are turned to glass. Geologic upheavals and global topographic changes transpire as coastlines change, rivers remove from their beds to carve out newer courses and some settlements disappear beneath the unstable earth. Civilization is fragmented into migrating groups, some traveling as far as distant America before further global changes from glacial melt water seas and quakes altered Earth again and prevented their return with impassable oceans. Sargon I of Akkad (Nimrod) was no more, his name by his people changed to Amraphel (He Made Us Fall), the renowned and celebrated Hammurabi. Interestingly, olden Hebraic texts read that during the catastrophe God *sent evil angels among* the people. These undoubtedly are the *Anunnaki*. Akkad was a world government in control of the entire ancient world, its collapse brought about the rise of multitudes of kingdoms and empires.

(395.4 years orbiting the sun)

1504 BC
(2391 AM)

No records; unseen from Earth.

(395.4 years orbiting the sun)

1109 BC
(2786 AM)

No records; unseen from Earth.

(395.4 years orbiting the sun)

713 BC
(3182 AM)

The Assyrian King Sennecherib led his army toward Jerusalem to subdue King Hezekiah of Judah after dispatching a letter to him boasting that there were no gods that could deliver Judah from Assyria. As the army approached, the Dark Satellite nearly collided into Earth and as in 1899 BC, an immensely powerful flux tube of electricity passed between the two celestial bodies, this time resulting with the complete vaporization of 185,000 Assyrian soldiers marching toward Jerusalem. This intense electromagnetic stress was further the cause of the entire planet's cessation of rotation and *retrograde* motion of 10 degrees, a phenomenon observed around the world as sundial's shadows moved in the *reverse*. As Earth hangs in space upon nothing, stabilized only by a magnetic field and locked into a gravitational orbit around the Sun, the Dark Satellite's mass and trajectory were enough to cause this temporary stop, reverse in the axial rotation by only 10 degrees, and then resumption of our planet's rotation. Never did the Earth stop orbiting the Sun, however, our planet was pushed about .1% distance further away from the Sun and the Earth's orbit became, in this year of 713 BC, more elliptical and was lengthened by 5.25 days from the original 360 days a year to 365.25. The 360 day year was the original span of the year according to the ancient Hebrews, Sanskrit and Brahmanic writings, Hindu texts, the Sumerian and Euphratean records, of Assyria, Egypt and Persia, these adding the Gatha Days, or 5 Bad Luck days which, on the other side of the world, were known as the Nemontemi, the 5 Useless Days of archaic Mexico.(1)  Pliny wrote that the Athenians previously knew the year to be 360 days (2) and amazing confirmation of this calendrical change comes from the fact that Roman records preserve that in the reign of King Numa Pompilius of Rome (this being his 3rd regnal year), he changed the calendar in Rome from 360 days to 365.25 days.(3)  The Thunderbolt (flux tube) that electrocuted the Assyrian army and the near collision of the Dark Satellite is further confirmed in the Annals of Lucius Piso (Book I) which relates that King Numa of Rome called lightning down from heaven through an incantation.(4)  King Hezekiah of Judah also altered the calendar.(5)  As attested in the legends and lore of cultures around the world, the Sun retraced its path in the heavens, or the Moon did so if the phenomenon was viewed on the other side of the world. In the Scriptures we read that the princes of Babylon sent emissaries to Jerusalem to consult the Temple records concerning ". . .the wonder which was done in the land. . ."(6) Some biblical translations read ". . .what happened to the earth."(7) The same event was mentioned by Greek historians to Alexander the Great, claiming that it was one of the ". . .great wonders recorded in their scientific books."(8) On the other side of earth the native Americans of North America reported that the Sun retarded its course by 80 hours (9) and in South America they claim that the Moon once fell from the sky and then later repositioned itself. (10)  Herodotus interviewed the Egyptian priests on this matter who claimed that the Sun had changed its usual position a few times in the remote past – "They assured me that Egypt was quite *unaffected* by this; the harvests and the produce of the river, were the same as usual, and there was no change in the incidence of disease or death."(11)  This was composed a few years prior to

440 BC. As we are now trespassing upon the domains of Immanuel Velikovsky, Zechariah Sitchin and other more competent scholarly sources on traditional facts concerning this strange alteration of the year to 365.25 days, we now return to this author's core thesis. The flux tube phenomenon was in modern times known in theory only, but when, in 1994 AD, the comet Shoemaker-Levy 9 approached Jupiter, the powerful lightning blast between the two bodies on a collision course was seen and photographed. The lesser of these bodies, the comet, was fragmented and broke into 21 pieces that slammed into Jupiter's surface, producing spectacular explosions and the release of tremendous energy. It was a similar flux tube that melted the bodies of the 185,000 Assyrians in the metal armor with their metal weapons – an expansive terrestrial *antenna*. In the *Apocalypse of Baruch* the text reads that (God Speaking) "And at that time I burned their bodies within, but their raiment and arms I preserved outwardly."(12)  The people of Judah later visited the spot and found the entire host had been vaporized in their armor. King Sennecherib, camping apart from the main host, survived and he and his entourage escaped back to Nineveh where he was assassinated by his own sons, who themselves escaped into exile.(13)  As the *evil* angels were released to afflict mankind in 1899 BC to aid in the destruction of those building the Tower of Shinar, now according to Hebraic writings like the *Apocalypse of Baruch*, holy angels led by Ramiel were unleashed by God to slay the Assyrians.(14)  In fact, the *Baruch* text is narrated by the angels that performed the deed.

The incredible synchronicity here lies with the fact that this happened at the conclusion of exactly six Mayan baktuns since the beginning of the *Mayan Long-Count Calendar* in 3113 BC – six baktuns of 144,000 days each, being precisely *864,000 days* (864: Foundation of Time number). Each Mayan baktun was 400 actual solar years of 360 days each, a fact the scholars are not unaware of, however, they dismiss this integral piece of information because the *present* length of the year in 365.25 days. The scholars are guilty of extreme arbitrariness, relegating the 13 baktuns (1,872,000 days) of the Long-Count on an entire system of 365.25 days to arrive at the ridiculous sum of 5125 years for the longevity of the Long-Count from 3113 BC, when the *original* system under the 360 day year equaled a perfect *5200 years*. This is how the scholars arrived at the impossible date of 2012 AD for the end of the Mayan Calendar, a date NOT significant to the Maya at all. The Long-Count Calendar is perfectly preserved in chronolitic geometry in none other than *Stonehenge*, a temple site erected by Enoch to demonstrate the awareness of the 360-day year and the fact that it would be broken *twice* in the future, at 713 BC and again at the END of the Mayan Long-Count in *2046 AD* when NIBIRU nearly *collides into Earth*, pushing our world closer to the Sun and *altering the calendar* once again. This is the 2400th year of the Long-Count, and from this 713 BC change in the year and orbital distance from the Sun, each baktun of 144,000 days now equals 395.4 solar years – the additional 5.25 days a year abbreviating the baktun of 400 to 395.4. The Mayan Long-Count system and Stonehenge will be reviewed in more detail later in this book. From this year forward remains 7 baktuns, or 1,008,000 days, which ends in 2046 AD. The Dark Satellite and Earth separate and the pattern continues.

(395 years of orbital period around Sun virtually mirroring a Mayan baktun of 144,000 days in post-713 BC system)

319 BC
(3576 AM)
Dark Satellite passes unseen from earth in synch with the conclusion of the 7th Mayan baktun, being 1,008,000 days into the system from 3113 BC.

(395 years orbiting the sun)

76 AD
(3970 AM)
This is an amazing synchronization astronomically and calendrically. This completed the 8th Mayan baktun, which began in 3113 BC (782 AM) of 144,000 days for a total of 1,152,000 days from its inception. The Dark Satellite passes over the ecliptic and is clearly seen from Earth as a great *javelin*, recorded by Titus. Pliny wrote that it was interpreted as an omen of doom.(15)  Intriguingly, this is the first year of the Saka Calendar of India.

(395 years orbiting the sun)

472 AD
(4366 AM)
The Dark Satellite is not seen in this year. As the orbital period is 395 years and the Mayan system is calculated at 144,000 days for each baktun, which measures at 395.4 years, the two are out of synch now by a two-year variance.

(395 years orbiting the sun)

867 AD
(4761 AM)
No records; Dark Satellite now three years variance from Mayan Long-Count.

(395 years orbiting the sun)

1262 AD
(5256 AM)
No records; Dark Satellite now four years out of synch with Mayan Long-Count.

(395 years orbiting the sun)

1657 AD
(5551 AM)
No records; Dark Satellite now five years out of synch with Mayan Long-Count.

(395 years orbiting sun)

2052 AD
(5946 AM)

This is the 10th pass of the Dark Satellite by Earth from 1899 BC when evil angels aided in dispersing mankind around the globe. And now, *six years* after the end of the Mayan Long-Count in 2046 AD when NIBIRU pushed earth toward the Sun and again *altered the calendar*, the Anunnaki, *evil angels*, are released from their lunar prison upon mankind to *gather the nations* they anciently dispersed into the Beast Kingdom that will be ruled by Abaddon, the *Eighth* Anunnaki King who fulfills the reign of the Seven Kings from before the Flood, whose reign were abruptly ended by the cataclysm. These kings are mentioned in the Book of Revelation. This amazing year is devastating upon humanity and is the subject of this author's work entitled *Descent of the Seven Kings: 2052 AD Return of the Fallen Ones*. The year 2052 AD is encoded within the Stonehenge II bluestone horseshoe and ring, as well as within the Great Pyramid's timeline (see *Chronotecture*) and within the recent mysterious *crop circles* (see *Chronicon*). This date marks the return of the Ancient Ones venerated in occult and satanic literature, the archaic MASKIM of Sumer, a race of demonic beings; of Thomas Burgoyne's Dark Satellite *inhabitants*, the haters of all mankind. Their physical descriptions are provided in vivid detail in Revelation 8 and conforms to Enoch's writings that declare that in the Secret Year ". . .Hell shall be opened," and come forth destroying.(16)  The Islamic Quran even reads that in the Last Days *angels* ". . .shall present Hell to unbelievers, *plain to view!*"(17)  This Islamic passage concerns the blowing of a Trumpet, and corresponding to this, the Anunnaki invasion of the *Seven Trumpets* is in Revelation 8.

Though this is a fascinating subject, our extant thesis concerns 2046 AD. The history of the Dark Satellite was provided because it is essential to understand how this lost moon in 713 BC altered the *solar year*, a fact that *no chronologist* known to this author has taken into consideration when factoring ancient dates and their relativity to modern times. The entire world has been duped into believing in the significance of 2012 AD, and for nothing.

## Archive 6

# Sodom-Trojan Apocalypse Comet Group

"God then cast a thunderbolt upon the city, and set it on fire, with its
inhabitants, and laid waste the entire country with like burning . . ."
—Flavius Josephus, *Antiquities*, in reference to
destruction of Sodom & Gomorrah (1.11.4)

Aside from the universality of the Deluge, there is no other body of traditions more prolific than the ancient and modern references to the fiery destruction of Sodom and Gomorrah. When concerning the wickedness or prosperity of nations, in the New Testament writings there is no greater lesson than that provided in the fate of these cities. But the judgment that came from heaven did not alone annihilate these cities in Canaan. The story of Sodom and her satellite cities begins with another civilization that also shared their fate. Elam.

The Elamites were ancestors of the Persians who inhabited the Indian subcontinent in the early 2nd millennium BC.(1)  This culture built advanced stone cities with canal works, sewage systems and forlorn cities have been found in the 20th century from underneath the desert wastes, now called Mohenjo-daro (Hill of the Dead), Harappa, Kalibanga and a unique port city named Lothal, which lies today many miles from a waterfront. Mohenjo-daro is spread out over a salt-encrusted desert (just like the Dead Sea region of ancient Sodom) but long ago it was a dense forest. The city probably housed about 40,000 people.(2)

In the year 1898 BC the Elamites, ruled by a former general under Sargon I of Akkad named Kudurlagamar (Chedorlaomer in Genesis 14) marched his army around Mesopotamia and forced Sodom and Gomorrah into paying an annual tribute of bitumen and oils – fossil fuel byproducts they harvested from the rich source of the Salt Sea (Dead Sea in Canaan). The bitumen was needed for building materials, lime mortar for bricks, glues, lamp oils and caulking ships.(3)  Taking auxiliaries from Canaan, the Elamite force then marched into the Nile Valley and subdued the Egyptians in the same way. That Sodom and Gomorrah, Admah and Zeboiim, four of the Cities of the Plain, were flourishing at this time is revealed by their mention in the 20th century discovery of the ancient Canaanite Ugaritic tablets.

In the year 1885 BC, after the passage of 13 years paying tribute to Elam, the Cities of the Plain rebelled and sent no more tribute. As the Elamites were engaged in a massive city-building scheme, they did not initially respond to this rebellion. Sargon I, now called Amraphel (Hammurabi/Ammurapi), interpreted this silence to be a weakness in Elam, and in the fifth year of the revolt of Sodom and Gomorrah, this Babylonian king, known also as *Nimrod*, assembled his forces to invade Elam. The Elamites amassed a lesser force, outnumbered 7 to 11, but both armies had among them many from the tribes of the Giants (biblical Rephaim, Anakim, Emims – the Titans of old). Nimrod brought his nephew Tudal, king of the Hittites, and his son Arioch's forces from Sumer. The conflict was immortalized in the annals of the ancient world as the famous *Battle of Kuruksata*, which saw the end of many of the Giants

49

(Nephilim). This was the Great War mentioned in many Euphratean texts, and Babylon lost. Nimrod, Tudal and Arioch were all forced to submit annual tribute of resources and auxiliaries to Elam. The conflict was over the right to control the Middle East's *fossil fuel resources*.

The Elamites lorded over Sodom and Gomorrah for 13 years. But now, 13 years after, they rebelled in the year 1872 BC (2023 AM). The Elamites under Kudurlagamar amassed a great army and were joined in a confederation by Nimrod, Tudal and Arioch. This multicultural invasion force marched into Aram (Syria) and laid waste the cities of the Giants in Bashan and Ashtaroth-Karnaim (Star of the Two Horns). The Rephaim, Emim and Zuzim giants all died or fled before them and even some of the Amorites in their path were killed. The attack from the north left the Cities of the Plain defenseless and the army defending Sodom and Gomorrah and their satellites were routed and chased off. The famous cities were sacked, and the women and children carried off as loot with all the merchandise and livestock. The Sodomites were left with nothing.

The biblical patriarch Abraham, a mighty lord with many servants, flocks, allies and wealth, summoned his friends. they included three Amorite chieftains and a host of 318 Anakim – the gigantic descendants of Noah and his wife through Anak (a giant born to them after the Flood). Among the captives is Lot, Abraham's nephew. The Elamite confederation was camped at Damascus, the ancient Baalbek, at leisure enjoying the spoils of drink, food and their enslaved women. With violence the small force led by Abraham attacked the encamped host and utterly slayed all who attempted to rally and oppose. The invaders were driven off and the entire trains of loot and captives were carried back to the cities of Sodom and Gomorrah. Kudurlagamar, Tudal and Arioch died, but in fulfillment to a prophecy known to Nimrod, he escaped back to Babylon. (see *Chronicon*).

This deliverance of the Cities of the Plain was the *cause* for their total and horrific extermination in 1848 BC. This rescue by Abraham began a 24-year period of peace and prosperity for Sodom and Gomorrah and the surrounding cities lived off the resources of the Dead Sea. The people were quickly ruined by excess. Their prosperity grew so rapidly they become haughty and isolationist. The old Hebraic records attest that these people began practicing hedonism, the tradition of three times a year openly trading their wives off and daughters. Homosexuality became rampant, trade was cut off to other nations because they had everything they needed, and foreigners and those poorer than the rest were persecuted, taken into custody, tortured publicly and executed in cruel and inventive methods in open courts. When foreigners attempted to pass through their domains all their merchandise was confiscated and spread out among the people and, due to complete judicial collapse, the judges of the courts invented clever ways to circumvent the law to the disadvantage of anyone bringing suit against citizens of those cities. Their lives degenerated into pure vice and luxury, and they spent their days contriving evil. Their wealth was incomparable and they serve as an historical example of how money is the root of all evil.

In 1848 BC Abraham's nephew Lot was visited by messengers (angels) from God. They warned him to get out of the city after the men of Sodom outside Lot's dwelling demanded to have the messengers inside delivered to them, intending to rape them (not knowing they were angels). Lot escaped with his daughters to a cave, but as the flaming destruction descended from the sky, Lot's wife was still in the fallout zone and through intense heat, radiation and the chemical fallout from the cometary detritus and powerful thunderbolts, her body was almost instantaneously *mineralized* into a pillar of salt. She became a *fossil*, an organic body that through chemical processes and intense heat and pressure petrifies into the minerals of the surrounding rock, which in this case was salt from the Salt Sea that was already covering the ground. Fossilization does not occur today, but in just a few more decades this turning of men into stone will transpire again on Earth.

As the cometary debris and detritus of the strewn field of the Sodom-Trojan Apocalypse Comet Group rained ruin upon the cities of Sodom, Gomorrah, Admah and Zeboiim, completely erasing this Canaanite civilization, the same fallout fell upon their enemies in India. The skeletons found in Mohenjo-daro and Harappa were discovered in contorted positions, often *holding hands* and left unburied, lying upon rooftops and in the streets.(4) This serves to reveal that these people *watched their doom approach*. Their skeletons are complete, which is unusual because natural carrion beasts would have devoured the corpses of those fallen in battle. It is a fact attested in many modern books that the Russian scientists studying these skeletons tested these bones to be 50 times more *radioactive* than the remains of those skeletons left behind in Japan after the detonations of the atomic weapons at Hiroshima and Nagasaki in 1945.

Because of these facts, some researchers have concocted fantastic stories of primordial nuclear exchanges between advanced civilizations long gone. But this radioactive contamination was by cometary fallout. The Indian epics these writers like to cite as evidence of atomic weapons long ago, like the *Mahabharata*, do on their surface seem to refer to nuclear weapons as darts or spears, however, it was the ancient habit of scribes to portray the fall of enemy cultures by cataclysms of quakes, meteoric rain and the like as the deliberate interaction of patron and local deities against their enemy cities. In essence, the epics were fictions brilliantly recording and adapting historical disasters for political means.

The proximity of large comets and bodies passing *through* our outer atmosphere but not entering totally for a collision would affect the breach of huge areas in the sky, virtual holes where the magnetic field would not repel the trillions of radioactive particles from the sun-windows in heaven raining toxic debris from space, killing all the inhabitants of the regions of the Indus Valley and Valley of Sodom. The epics read that intense burning and melting afflicted these cities, incendiary clouds and terrible winds. Our magnetic field repels enough harmful solar radiation as to keep the planet's inhabitants safe from contamination, but a regional weakening of the field would result in bad sunburns, fiery windstorms or even geologic upheavals.(5)

The intense heat is evidenced by the presence at both sites of countless tektites – strange smoothly-melted stones thought to be of meteoric origin.(6) These odd rocks are found at other sites around the world, all these areas having in common the presence of seemingly dateless ruins abandoned long ago, structures having been vitrified (changed to glass by intense heat).(7) Further scholarly evidence that the Indus Valley civilization of Elam met the same fate at the same time as Sodom and Gomorrah is found in the fact that these academics and scientists believe these forgotten cities all came to an end in approximately 1900 BC (8), and this *approximate* date happens to align perfectly with the actual 1848 BC event by only 52 years.

The Lamentation Texts of Sumer reveal that the destruction was linked to the evil cities of the west that met the same ruin. The Lamentation Texts reveal that the immense urban centers of Sumer were laid to waste – Ur, Uruk, Nippur, Eridu and others – with dead bodies laying about in heaps, children with mothers, fires ablaze, an Evil Wind, all parts of the disaster according to Sitchin.(9) Though southern Mesopotamia was ravaged by the storm, Babylon was actually *empowered* by an influx of refugees. In fact, among those that added to Babylon's ranks were *Elamites*, according to Sitchin.

The Arabic lore concerning the date of these cities was penned in the 7th century AD by Islamic scribes and included in the Quran. They wrote that a fierce, roaring wind ". . .which He (God: Allah) imposed upon them for seven long nights and eight long days, so that you might have seen men lying overthrown, as if they were hollow trunks of palm trees."(10) Another intriguing passage in the Muslim writings seems to confirm the archeological fact that skeletons were found in the streets holding hands. ". . .then when *they beheld it* (their judgment coming), a dense cloud coming toward their valleys, they said, 'Here is a cloud bringing rain,'. . . Nay, but it is that which you did seek to hasten, a wind wherein is painful torment destroying all things. . . morning found them so that nothing could be seen save their dwellings. . ."(11) Even the Roman historian Tacitus relates that at the edge of the Dead Sea were once great cities that were totally destroyed along with the fertility of the ground by thunderbolts which burned the earth.(12) As seen by our caption heading this archive, this is also mentioned by Josephus.

We have devoted particular attention to the end of Sodom and Gomorrah and the Elamites because these histories were included in the body of archaic scriptures for our *future* instruction, the mold of what is to take shape in years to come, fashioned within the framework of the years in antiquity. But before we get to this fascinating study of the future, we must review the Sodom-Trojan Apocalypse Comet Group. Seven years (2520 days: 360 x 7) after NIBIRU exited the inner system, passing over the ecliptic in 1855 BC, an expansive strewn field of cometary detritus and asteroids trailing the Anunnaki Homeworld broke free of NIBIRU and entered its own solar orbit, beginning to pass Earth in the *2046th year* of the Annus Mundi Chronology, or 1849 BC.

| | |
|---|---|
| 1849 BC (2046 AM) | Comet group begins passing over ecliptic in two-year train through 1848 BC. |
| 1848 BC (2047 AM) | The cities of Sodom, Gomorrah, Admah and Zeboiim in Canaan along the Salt Sea (Dead Sea) are literally fossilized as they are collapsed by quakes and burned under intense pressure. The cities of the Elamites in the Indus Valley meet the same fate, though many survivors relocate to the north and west, later becoming a sizeable ethnic element in Persia. |

(15 years orbiting over ecliptic)

| | |
|---|---|
| 1834 BC (2061 AM) | No records. |

(286 years orbiting under ecliptic)

| | |
|---|---|
| 1548 BC (2347 AM) | No records. |

(15 years orbiting over ecliptic)

| | |
|---|---|
| 1533 BC (2362 AM) | A comet is seen brilliantly in the heavens by the Chinese and the Egyptians. According to the Tulli Papyrus, the priests of Karnak (Upper Egypt) recorded that in the 22nd year of the reign of Thutmose III there appeared a very bright comet (disk of fire), one very different that those known of old.(13) The papyrus describes that a comet broke apart in the sky and became many smaller disks, this spectacle filling the heavens.(14) Interestingly, this is 366 years (half of 792) after the Babel catastrophe of 1899 BC. |

(286 years under ecliptic)

| | |
|---|---|
| 1247 BC (2648 AM) | Comets enter inner system and begin passing over ecliptic in a 3-year long train of debris to 1244 BC. |

**1244 BC**
**(2651 AM)**

A bright comet appears in the sky and is recorded in the annals of King Shalmaneser I of Assyria as appearing during a battle between the Assyrians and Hittites.(15)  This comet seemingly presaged the impending ruin of the Hittite Empire. This same comet served as a key element is the oldest Greek historical memory known, the battle known as *The Seven Against Thebes*, when seven early Grecian nations besieged the Greek city of Thebes over a domestic dispute and were humiliatingly beaten back. They embarked upon the road to battle despite the unfavorable omens and *signs in the heavens*.(16)  The failure was a source of ridicule in the soon-following Trojan War. The comet of 1244 BC began a countdown of 10,800 "evenings and mornings," until the end of Hittite civilization. While the Sodom-Trojan group orbited over the ecliptic for 15 years, in the year 1239 BC King Agamemnon of Mycenaea began his siege against the city of Ilium, known as Troy, initiating the famous Trojan War. The conflict would last a decade, resulting in the sack of *many* cities in Asia Minor belonging to the Trojans.(*Iliad*: Book I, lines 184-188). (17)  The Trojan War began exactly in the 1000th year after the Great Flood (2239 BC).

(15 years orbiting over ecliptic)

**1232 BC**
**(2663 AM)**

As the train of three years enters the inner system in this the 7th year of the Trojan War, according to Homer's *Iliad*, a plague afflicted the Achaeans under Agamemnon.(18)

**1229 BC**
**(2666 AM)**

This is the 10th year of the Trojan War and the *366th* year (half of 792) of the Hittite Empire seated at *Babylon*, which was founded in 1595 BC. Asia Minor, where the Hittite civilization flourished (Hittite Dynasty of Babylon succeeded the Amorite Dynasty at Babylon) fell victim to a storm of thunderbolts from space during the daytime that vitrified entire towers, citadels, stone cities and burned the people and their dwellings. Hittite Anatolia was reduced to ashes and glass, as were the Indus and Sodomite valley cultures before them. The unfolding of the famous Trojan War was along the coast of Asia Minor west of Anatolia, and these unfortunate regions did not escape this cometary destruction. As this was the 10th and final year of the conflict, Agamemnon invaded the plain of Troy with a force of 60,000 warriors from the Greek mainland with over 20 proto-Greek, Aegean, Mediterranean and Greek subcultures in a fleet of 1186 vessels.(19)  Allied to King Priam of Troy and present at the conflict were the soldiers of nations from deeper Asia Minor, the Mediterranean and Asia. Amazingly, Phrygia, Caria and Lycia were involved in the war and these nations in Asia Minor were garrisoned with *Hittite* troops for Ilium, where Troy was seated, and the coastal city-states of Asia Minor along the Aegean, were a buffer zone between Hittite Anatolia and the emerging might of Mycenaea. The astronomical chaos that ended the Hittites did not escape unnoticed in Homer's epic texts. In this year, according to the *Iliad*, a *comet* with a long tail was seen over Troy, said by Homer to be a sign from God.(20)  During the day, with the Sun shining brilliantly, terrible *lightning bolts* from heaven struck the Greek army in front of the walls of Troy, hitting two different areas of the battlefield (possibly seen all around, as lightning struck surrounding regions).(21)  Also during the battle it *rained blood* and a destructive earthquake occurred.(22)  Some accounts hold that fiery arrows fell from the sky.(23)  Many Trojans escaped the war and environmental chaos by sea and, en masse, colonized a region of Italy, later integrating with a Semitic culture known as the Etruscans. After several battles with Latins and other Etruscan groups, these descendants of Troy built a city and called it *Rome* (see *Chronicon* 753 BC).

(286 years under ecliptic)

946 BC          No records.
(2949 AM)

          (15 years over ecliptic)

931 BC          King Solomon died and the Monarchy of the three kings of Saul, David and
(2964 AM)       Solomon ended with the political and national division between Judah and
                Israel beginning the *Divided Kingdom Chronology*. Israel would be deported
                by Assyria and lost into the cultures of Asia Minor, Persia, Media, Babylon,
                Syria, Armenia, deeper Asia and ultimately into the diversified European
                nations, but Judah would remain intact until AD 70 when the *Romans*
                destroyed Jerusalem and the Temple, and then the Great Diaspora of 135 AD,
                when Rome attempted to exterminate the Jews. No records of a comet.

          (286 years below ecliptic)

645 BC          The train enters the inner system in a 3-year debris train to 642 BC.
(3250 AM)

642 BC          Just after the Romans defeated the Sabines at Mantrap Wood a *rain of stones*
(3253 AM)       descended upon Alba Mount near the city of Rome. Reports filtered in that
                the peculiar rain had afflicted the countryside from out of a clear sky. Men
                were dispatched to investigate the stones and shortly afterward a *plague* broke
                out that even sickened King Tullus Hostilius. Then a lightning bolt struck the
                Palace in Rome and burned up the structure, killing the king.(24)  This event
                parallels exactly the event of 1229 BC, the 3rd year of the Sodom-Trojan
                group, as meteorites and strange lightning occurred.

          (15 years orbiting over ecliptic)

630 BC          No records. The strewn field at this time, probably due to loss of mass,
(3265 AM)       deteriorates in orbital velocity by five years in its sub-solar orbit of 286 years
                to 281. This lengthens the train to about 12 years.

          (281 years orbiting below ecliptic)

349 BC          In this the 3rd year of the 107th Olympiad, as King Philip (renowned father
(3546 AM)       of Alexander of Macedon) was causing chaos throughout Greece, the sky
                appeared to open and *blood and fire* rained upon the land.(25)  Alexander the
                Great was seven years old.

          (15 years over the ecliptic)

334 BC
(3561 AM)   Alexander the Great inherits his father Philips' elite fighting forces and kingdom, fulfilling biblical prophecy foretold by Daniel concerning a 2300-year period to a date in the far future when the Temple Mount would be consecrated back to the Jews. This 2300 years, beginning with the emergence of Alexander, ended in 1967 AD when Israel took back the Temple Mount from the Muslims in the Six Day War (see *Chronicon* for details). The train's 281-year sub-solar orbit now decreases to 280 years.

(280 years orbiting below ecliptic)

54 BC
(3841 AM)   Pliny wrote that in this year *iron* rained from the sky in Lucania, and was a bad omen, for the following year the famous Roman General Marcus Crassus was killed in battle with the Parthians. In his ranks were many Lucanian soldiers. The iron was spongy in shape.(26)

49 BC
(3846 AM)   During the Civil War between Pompey and Caesar a bright comet appeared. (27)  Caesar prevails and many believe his first year, 48 BC, was the end of the Republic.

44 BC
(3851 AM)   As the force under Marc Antony seiged the army of Decius Brutus at Mutina, a storm of shooting stars appeared and Pliny recorded that a *second Sun* appeared in the sky with our own Sun.(28)  In a conspiracy, the Senators assassinate Julius Caesar and to the astonishment of the world, the *Sun darkens* for an unusually long time (comet transiting occulting the Sun).(29)

43 BC
(3852 AM)   In this year the Second Triumvirate was formed between the military leaders Marcus Antony (famous lover of Cleopatra), Marcus Lepidus and Gaius Octavius (Octavian), and a *comet* was viewed in the western heavens.(30)

42 BC
(3853 AM)   Agents of the Triumvirate hunt down and kill the prime conspirators that orchestrated Julius Caesar's assassination and a *strange Sun* appears with our own.(31)  This is the final comet of the Sodom-Trojan Apocalyspe Comet Group. The comets that looked like suns to the Romans were in fact *disintegrating* comets, objects fracturing and fragmenting into millions of pieces, as happened to Shoemaker-Levy 9 when it approached Jupiter in 1994.

This comet group began with NIBIRU, ending the civilizations of Sodom and Gomorrah, the Indus Harappan culture and the Hittite Civilization of Asia Minor. It was partially responsible for the Fall of Troy and heavily afflicted the Romans, who were descended from these Trojan heroes. These historic examples provide us prophetic foreshadowings. The Hittites were an Iron Empire, followed in the western migration of empires by Rome, the Iron Empire in Daniel's biblical prophecies. The ancestors of Rome were an offshoot of the Hittites, preserved in the Latin *Italic*. And as we will find, the descendants of Rome, a *modern* Iron Empire, the United States of America, will fulfill these types and shadows perfectly.

## Archive 7

# Romanid Apocalypse Comet Group

It is a long established fact that plagues follow the eclipses
of the Sun.
— Pliny, *Natural History*, Mining and Minerals (1)

As the Anunnaki Homeworld traverses the inner system in 300 BC, a glacial mass in its train breaks up into comets, ice-encased asteroids and huge-to-small, gravel-sized stones that begin their own orbit in a 20-year train of debris free of NIBIRU's gravitational hold. They do not maintain the velocity of NIBIRU and exactly 108 years after entering its own solar orbit, the Romanid Group enters a stable, 136-year orbit after passing close by the Earth. Like a long snake traversing the ecliptic, Earth literally moves *through* the detritus train every year that this group is crossing the ecliptic plane. Called the Romanid Apocalypse Group because it was the Romans at these times that were an international power and it was from their observations that astronomical records were left to us.

| | |
|---|---|
| 222 BC<br>(3673 AM) | Train of 20 years begins. During the consulship of Gnaeus Domitius and Gaius Fannius it was observed that *two moons* appeared in the sky with our own Moon, as recorded by Pliny.(2)  For objects to be equated with our Moon denotes that these were small planetoids or moons once trailing NIBIRU, or *gigantic* asteroids our world passed as we moved along the ecliptic around the Sun at 1.3 million miles per hour. |
| 217 BC<br>(3678 AM) | This was during the Second Punic War between Rome and Carthage. So many earthquakes occurred throughout the world that 57 of them alone were felt at Rome.(3)  This was 5th year of the Romanid Group. |
| 215 BC<br>(3680 AM) | During this year in the Second Punic War it was recorded to have *rained blood.* (4)  This had occurred at the Exodus over Egypt and later again in the Trojan War, the ancestors of the Romans. As will be seen in this work, blood would fall from the sky again in the future and beyond into the Apocalypse. |

207 BC
(3688 AM)
A famine kills myriads in China, lasting to 204 BC. King Shih-Huangi-ti of the Ch'in Dynasty died and the various Provinces reverted back to anarchy and feudal struggles. The *Irish Annals* record that their cattle died off in great numbers this year. In this famous year, the Romans in the Second Punic War defeated Carthaginian General Hannibal's brother, Hasdrubel, in the Battle of Metaurus.

206 BC
(3689 AM)
Roman historian Livy wrote that in this year appeared *two suns* over Rome, (5) their brightness hinting that they were comets with trailing debris tails pointing toward Earth and reflecting sunlight.

205 BC
(3690 AM)
Roman historians recorded that stones fell from the sky. (6)

204 BC
(3691 AM)
Odd lights are reported in the night skies. Famine finally subsides in China.

(136 years orbiting the sun)

86 BC
(3809 AM)
Train of 20 years begins. In this year of the consulship of Lucius Valerius and Gaius Marius, an object like a *giant blazing shield* traversed across the sky from east to west, toward sunset, giving off sparks as it moved.(7)  If the object impacted, it did so too far away from Rome for news of the disaster to be reported. Many similar objects have been seen throughout history of things passing through the upper atmosphere but never quite entering before disappearing back into space.

66 BC
(3829 AM)
It appears that Earth only passed through the first and last years of this train of debris without encountering anything significant. In this year, during the consulship of Gnaeus Octavius and Gaius Scribonius, a *strange star* appeared moving in the heavens. A spark was seen to have come from this star, and the spark grew in size (approached the Earth) until it was the *same size as the Moon*. It appeared to be enshrouded in clouds (had an atmosphere?) and then departed as if on fire. This was recorded by Pliny, and was noted also by the Proconsul Silenus.(8)

(136 years orbiting sun)

51 AD
(3945 AM)
Train of 20 years begins. At this time, during Claudius' consulship with Cornelius Orfitus, a strange sun appeared in the sky in addition to our own, according to Pliny.(9)  This was probably a disintegrating comet almost in transit. Tacitus wrote that in this year, repeated earthquakes afflicted Rome and the countryside.(10)

53 AD
(3947 AM)
Earthquakes afflicted Phrygia in Asia Minor.(11) In this year the people of Ilium petitioned Emperor Nero to exempt them from all public burdens solely on the fact that Rome descended from Troy (city of Ilium).(12) Interestingly, it was during the Trojan War that earthquakes afflicted Phrygia.

56 AD
(3950 AM)
Unusual lightning struck both the temples of Jupiter and Minerva in Rome.(13) Under normal circumstances this could be relegated as coincidence, but the following effects of the Romanid Group dispels this, for the lightning is of an extraterrestrial nature.

58 AD
(3952 AM)
In this the 7th year of the passing of Earth through the train of Romanid debris, as the Roman army destroyed the city of Artaxata in Armenia, a divine portent occurred, according to Tacitus. "While the Sun shone brilliantly all around the walls, the area of the city itself was suddenly enveloped in a dark cloud with unearthly lightning flashes."(14) Also, an ancient fig tree named Ruminalis, believed to have been planted in the days of Rome's first king, was already about 830 years old when it shriveled up and seemingly died. Its trunk withered and shoots died. But after a short while the tree revived and grew fresh shoots.(15) This was the same year the apostle Paul was confined by the Roman Governor Felix in Caesaria.

59 AD
(3953 AM)
In this the 8th year of the Romanid Apocalypse Comet Group, the famous city of Laodicea was destroyed by an earthquake and during the day, the Sun darkened over Rome, according to Tacitus. All 14 Districts of Rome were struck by a fierce lightning storm, one woman having been killed by a thunderbolt while in her husband's arms.(16) The Romans recorded the appearance of a comet about the same time that a flash of lightning struck and broke the table Nero was eating at in his mansion at Sublaqueum.(17)

62 AD
(3956 AM)
A violent earthquake damaged buildings at the luxury-cities of Pompeii and Herculaneum near Mount Vesuvius in Italy on February 5th.(18)

63 AD
(3957 AM)
Lightning struck the Gymnasium in Rome, burning it down, melting a large bronze statute of Nero into a shapeless mass.(19)

64 AD
(3958 AM)
In this year lightning struck a 120-foot high portrait of Nero, the largest painting ever. The portrait was on linen and immediately burnt up in the gardens of Maius.(20) A comet was recorded at year's closing along with an unprecedented display of lightning, accompanied with abnormal birth defects. (21) The City of Rome, seat of the empire, burned for *seven days*. While many believe Nero intentionally set the fires, the evidence suggests that the fires were ignited from extraterrestrial sources (lightning or incendiary fallout), and Nero's gross negligence in organizing an effort to staunch the raging fires was probably because he had a dream to rebuild Rome according to his own specifications. Tacitus writes that after the fire, the general rumor among the people was that Nero was witnessed singing a hymn of the Fall of Troy as the city burned. This was the *108th year* since the start of the Roman Julian calendar.

65 AD
(3959 AM)    Earthquake in Ireland sank the entire region of the coast and hinterland into the sea, according to Girgaldus Cambrensis.(22)

66 AD
(3960 AM)    Flavius Josephus wrote that a comet appeared. The city of Rome itself was decimated by a plague, filling the houses with corpses.(23) Judea began to riot and rebel against Rome and even succeeded in killing an entire Legion under General Cestius Gallus, initiating the Jewish War. It is Josephus who led the Jews against the Romans. Josephus' comet appeared like a *sword* (comet tail), as well as another object (perhaps asteroid). It was attended by an *earthquake*. (24)

68 AD
(3962 AM)    In this the 17th year of the Romanid Apocalypse Comet Group an unusual occurrence of geological displacement transpired, recorded by Pliny. An entire Roman estate was completely *turned around*.(25)  In this year the Senate declared Nero to be an Enemy of the Public and sentenced him to suffer a slave's death, to be whipped until dead, but he commits suicide instead.

69 AD
(3963 AM)    Roman General Vespasian offered sacrifices upon Mount Carmel at an ancient altar site in Israel (Judea), mentioned by Tacitus. The local priest prophesied that Vespasian was destined for greatness, and astrologers confirmed this by noting strange omens in the heavens.(26)  In this same year Vespasian became Emperor of Rome and his son, the General Titus, was appointed to put down the Jews.

70 AD
(3964 AM)    Earth tremors shook the Temple on the day of Pentecost and voices loudly pronounced from the Temple Precinct, "LET US REMOVE HENCE," striking terror into the hearts of those that heard this.(27)  Multitudes of witnesses see what they believe are hosts of supernatural beings battling in the sky with fiery weapons and a strange cloud illuminates the Temple. In this, the 1000th year since Israel and Judah separated into the Two Kingdoms in 931 BC, Jerusalem is totally destroyed by the Romans. Over a million Jews were killed in the four-year war and 97,000 more were sold into the slave markets. The fields were salted and the Temple treasury and library were sacked, the famous objects prominently displayed upon the Arch of Titus in Rome.

This 20-year train will continue as a long chain of debris spanning 20 years, however, its orbital dynamics are altered. The original 136 year orbit is now 153.5 years, the Romanid Group slowing down by a factor of about 17.5 years as they orbit the Sun.

(153.5 years orbiting the sun)

204 AD
(4097 AM)    No records for entire 20 years.

(153.5 years orbiting the sun)

| | |
|---|---|
| 358 AD<br>(4252 AM) | The 20-year train begins. Terrible earthquakes shook the cities of Macedonia, Asia, Pontus, the metropolis of Bithynia (Nicomedia) received the worst fate. At dawn on August 24th the clear sky quickly gave way to thick masses of black clouds that descended to the ground, blotting out the Sun and casting the whole region into blackness. Ammianus Marcellinus reports that the destruction began from the sky in the form of loud crashes, winds, fatal thunderbolts and a torrential downpour. An earthquake accompanied all this. Just as fast as it had begun, the quaking ceased, the sky cleared and a beautiful day was ruined by the visage of thousands of torn bodies, some still moaning among the piles of debris that once was a thriving city. Bodies of the dead and those still suffering were seen transfixed, impaled upon sharp timbers while others could be heard screaming beneath the rubble. Many people were trapped in the lower basements and cellars, recesses under the city. A fire quickly spread through the ruins for five days and nights, killing off all the survivors who were unable to get out of the rubble.(28) |
| 359 AD<br>(4253 AM) | At the city of Antioch, the capitol of Syria on the Orontes River, one of the four major cities of the Roman world (called the Jewel of the East), in a suburb known as Daphne, a child was born with two heads, a beard and very small ears, the populace regarding these birth defects as the birth of a monster.(29) May have been caused by the influence of the Romanid Group passing. |
| 362 AD<br>(4256 AM) | In the 5th year of the passing of the Romanid Group, on December 2nd, an earthquake finished off the city of Nicomedia, destroying what little remained or had been rebuilt after the terrible cataclysm of 358 AD. Much of Nicea was also ruined, the two quakes being 52 months apart.(30) |
| 363 AD<br>(4257 AM) | Several comets were seen at the same time, some in broad daylight.(31) Constantinople (ancient Byzantium) was shaken by an earthquake and Julius Augustus witnessed a shooting star of unusual brilliance.(32) The Etruscan diviners interpreted the omen as bad, using the authoritative Tarquitian Books (last official consultation of the Sybilline Prophecies), for anyone thinking to invade another's realm. Nonetheless, Julian later that year marched his army toward Persia with disastrous results to the army, resulting in his own death. But prior to this, in this same year, architects and agents of Julian arrive in Jerusalem with the intent of *rebuilding the Temple*, but strange fire-bursts burned the laborers to death, a phenomenon that happened a few times before the work was abandoned.(33) |

364 AD
(4258 AM)
A general madness seems to have blanketed mankind in this year. The entire world appears to be at war. Persia invaded Armenia and German tribes invade many Roman provinces. The Alamanni ravaged Gaul and Raetia, the Sarmatians and Quadi were laying waste to Pannonia and Britain suffered onslaughts from the Picts, Scots, Saxons and the Attacotti. Goths invaded Thrace and Moesia, while in the southern provinces of Africa the Moors and Austoriani were raiding Roman settlements and their allies. All the while, Romans were fighting each other. One wonders if this phenomena was not the direct result of the displeasure of the Deity, for in this year the Roman Church at the Council of Laodicea ordered that the Christian world would henceforth observe not the Sabbath (Saturday), but every week the *first* day of the cycle, violating the ancient Ten Commandments: "Remember the Sabbath and keep it holy."

365 AD
(4259 AM)
On July 21st an earthquake shook Alexandria, Egypt. This was in the 8th year of the Romanid Apocalypse Comet Group. Roman historian Ammianus Marcellinus wrote that fierce lightning and thunder, which continued through the quaking, preceded the quake. The sea outside the port withdrew far away from the shore and the inhabitants of the city found themselves far inland, looking out over an immense drying region of stranded fish and shelled creatures trapped in mud, vast valleys and mountains where sea had been before. As a great throng of people went out to gather the stranded sea animals, a tsunami came rushing back at hundreds of miles per hour overtaking them. The wave and returning sea rushed up the coast and drowned the city of Alexandria, even depositing ships atop buildings. When the waters subsided and drained from the city and hinterlands, many people were pulled out to sea to never be seen again. Approximately 50,000 people died.(34)  In his *History of the Later Roman Empire*, Marcellinus relates that this disaster overwhelmed the whole world, surpassing anything related either in legend or authentic history (his opinion). He wrote that the quake was felt from Egypt into the Aegean and that towns and buildings on coasts were destroyed when the sea was removed from its bed. Even in the Aegean, the returning sea deposited vessels far inland, some still holding their dead crews. Near Mothone in the Pelopponesse, he personally surveyed such a wrecked ship. Marcellinus wrote that ". . .the whole face of the earth was changed by a mad conflict of the elements."(35)  The lightning storm before and during the quake demonstrates the unusualness of this disaster.

368 AD
(4262 AM)

A comet appeared shining brighter than the Sun, according to Cyrillus to Constantine, Constantine the Great's son.(36) This is 11th year of the Romanid Group.

375 AD
(4269 AM)

In 18th year of the Earth passing through the Romanid Group, several comets were seen blazing in the sky, according to Ammianus Marcellinus.(37) The men of the day believed that these comets were warning Rome of impending dangers. And as history revealed, the Huns invaded the following year and forever changed the demographics of Europe.

As with the other comet groups, the Romanid train disappears from history only in that its orbital chronology seems to end. The debris does not vanish, however. As we will review later in this work, in the past few centuries from about the mid-16th century until now, there is a tremendous amount of icy and rocky objects orbiting the Sun that are so numerous (and thoroughly ignored by scientists) that no distinguishable patterns can be detected. The larger groups stemming from NIBIRU are definable, but as time and entropy take their toll, the refuse left behind by these groups literally *fills* the inner system.

## Archive 8

## 2046 AD NIBIRU Comet Orbit
## and the 2047 AD Reuben Comet Group

And the world was terrified, and many
cities of the nations fell in that same hour.
The earth did groan and shake mightily by
the power of the star of ruin.
— *Apocalypse of Matthias* 5:27

Earlier in this work we reviewed the orbital history of the Anunnaki Homeworld and learned that the only year in history when Both NIBIRU *and* planet Phoenix were in the inner system at the same time was 522 AD. NIBIRU was crossing the ecliptic after completing its 792-year orbit at the exact same time Phoenix was passing over the ecliptic ending its 138-year orbit. This was the year 5760 of the Anunnaki Chronology, which began in 5239 BC (see *Descent of the Seven Kings*), this being 144 x 40 or 360 x 16.

The train of NIBIRU has continually seeded this solar system with debris and this pass is no different. An immense strewn field breaks away from NIBIRU in this year and becomes the 2046 AD NIBIRU Comet Orbit group, while another one enters its own orbit around the Sun, as the 2047 AD Reuben Comet Group. The Anunnaki planet is fracturing and literally disintegrating. Its next pass in the 13th century AD will demonstrate this profoundly.

The 2046 AD NIBIRU Comet Orbit group stabilized into their own 434-year orbit around the Sun after 222 years earlier breaking away in 522 AD. Their orbital cycle will remain fixed at 120 years under the ecliptic and 314 years over it. This unique 434-year orbital chronology is referred to as an epoch of *Judged Time*, the author Stephen Jones in his work entitled *Secrets of Time*, exhibits evidence that many historical events are separated by cycles of 434 years or multiples thereof.

## 2046 AD NIBIRU Comet Orbit

| | |
|---|---|
| 742-748 AD (4636-4642 AM) | Strewn train enters the inner system in a 6-year long field. |

| | |
|---|---|
| 743 AD (4637 AM) | The *Irish Book of the Four Masters* records that strange stars appeared in the sky.(1) |

744 AD (4638 AM)

The *Irish Book of the Four Masters* records that there was seen numerous stars that fell from the sky.(2)  As this group is made up of *two large comets* and much smaller detritus, and this Anno Domini year in 744 (186 x 4 faces of Great Pyramid at its summit: see *Chronotecture*), we find in Hebrew the gematria for "two great lights," is 744. Additionally, this year is the 1488th year of the Post-Exilic Chronology, or 744 + 744 (1488). This chronology began in 745 BC when the Assyrians deported the Ten Tribes of Israel, which later migrated into Asia and Europe, descendants of these Israelite tribes settling in *Ireland*.

862-868 AD (4756-4762 AM)

Earlier in this study was revealed how the sum of 864 (144 x 6) is the *Foundation of Time* number linking seconds, minutes, hours, days, months and years. Amazingly, this date is linked herein to the Great Pyramid measurement of 744 by the period of 120 years, the *foundation* number (of the Anunnaki NER system, explained in *Descent of the Seven Kings*). For example, Noah was warned the Earth would be destroyed 120 years later. Foundations imply the *beginning* of something (all architectural projects require a foundation to begin). And this is precisely what we find. In 864 AD occurred, according to the *Irish Annals*, *both* a lunar and solar eclipse.(3)  But this is so rare as to be unbelievable, and astronomers note that no eclipses occurred over Ireland in this year. We see here how this Sun darkening was due to comet transit occulting the Sun of the 2046 AD NIBIRU Comet Orbit group. Amazingly, this EXACT year completes the 10th baktun of the *Mayan Long-Count* for a total of 1,440,000 days since the system began in 3113 BC. This is truly a *foundational year*, for it begins a countdown of 3 Mayan baktuns of 144,000 days each (432,000 days) to the *comet impact* and appearance of NIBIRU in *2046 AD*, while also beginning 3 Cursed Earth periods of 414 years each (see *When the Sun Darkens* for Cursed Earth system demonstrated) for a total of 1242 years to the year 6000 Annus Mundi (2106 AD), Armageddon, and the *return* of NIBIRU after orbiting the Sun for 60 years, since last passing earth in 2046.

(314 years orbiting the Sun over ecliptic)

1176-1182 AD
(5070-5076 AM)

A monk named Gervase in 1178 AD recorded that at Canterbury, England, monks watched in amazement at the clearly visible *crescent Moon split in two*, then set on fire like a torch. After the fire went out the Moon went black on June 25th.(4)  This was the collision of one of the two principle comets of the 2046 AD NIBIRU Comet Orbit group into the surface of the Moon, the impact releasing a blast equivalent to many atomic weapons exactly 434 years after 744 AD.

(120 years orbiting under ecliptic)

1296-1302 AD
(5190-5196 AM)

In 1296 AD King Edward I relocates the Stone of Destiny, also known as Lia Fail, from Scotland to London, England. This was the ancestral Israelite relic from the days of Jacob brought to ancient Ireland by Jeremiah the prophet in 583 BC after the fall of Jerusalem to Babylon in 585 BC. Jeremiah also brought Baruch the scribe and Jewish princesses. The Stone had been in Scotland for exactly *792 years* (NIBIRU's orbit), and like the pre-flood Anunnaki, the Stone was a symbol of *kingship*. King Edward I had the Stone set into the *foundation* of the Coronation Chair at Westminster Abbey where it remains today.(5) See also *Chronicon*.

1298 AD
(5192 AM)

This begins a period of 16 years of widely reported earthquakes, comets, tsunamis, displacement of lakes and rivers, mysterious astronomical signs, plague, fogs and general disasters from China, India, the Middle East, to Africa, Asia Minor, and Europe.(6)  This history is so established that it becomes redundant to recall it again here.

(314 years orbiting over the ecliptic)

1610-1616 AD
(5504-5510 AM)

The 6-year train of debris begins entering the inner system. There are no known unusual astronomical records for this period.

(120 years orbiting under the ecliptic)

1730-1736 AD
(5654-5660 AM)

The 6-year train begins entering the inner system.

1730 AD
(5654 AM)

Earthquake activity begins in the Canary Islands and will continue for five years until 1735 AD. These quakes release noxious gases that kill much wildlife.(7)  A quake at Hokkaido, Japan kills 137,000 people.

1731 AD
(5655 AM)

From Florence is witnessed a luminous cloud moving at a high velocity in the sky which disappears over the horizon, attended with quaking on December 9th.(8)

| | |
|---|---|
| 1732 AD (5656 AM) | On May 28th over Swabia are seen thick mists in the sky through which was visible a dim light and globes of fire in the air, all preceding an earthquake.(9) |
| 1733 AD (5657 AM) | Influenza (from *influence*-of the stars) sweeps through New York and Philadelphia in the British Colonies of America. Interestingly, this year is the admission of the 13th Colony, Georgia, which is exactly *1776 years* after the start of the Roman Julian Calendar in 45 BC – of course, 1776 AD being the year the 13 Colonies became the United States of America. |
| 1737 AD (5658 AM) | Though this is a year after the departure of the group detailed here, in this year occurred a terrible quake at Calcutta, India that killed 300,000 people. In this year over Carpentras, France, it rained soil and earth from the sky on October 18th.(10) |

(314 years orbiting over the ecliptic)

| | |
|---|---|
| 2042-2048 AD (5938-5942 AM) | The 6-year train begins entering the system. Earth in 2042 AD will already suffer from fallout of the King of Israel Great Orbit group, and the asteroids of this group may become meteorites afflicting the planet as well. Earth will experience all the attendant phenomena of these two groups from 2042-2046 AD, the IMPACT of a comet of the 2046 AD NIBIRU Comet orbit occurring just prior to the arrival of NIBIRU in 2046 AD. The year 2046 AD will be covered later in this work. This train does not end until 2048 AD and, as we will find, the apocalyptic destruction will continue to rain upon our planet even at this time. The 2046 AD impact is precisely 868 years after its sister comet collided into the Moon in 1178 AD, or two Judged Time epochs of 434 years each (868). |

## 2047 AD Reuben Comet Group

This group is called the Reuben group because of its apparent theme relating to the French and Italians, and by extension the Normans descended from France, all peoples having some ancestral connection to the Israelite tribe of Reuben that migrated with the other lost tribes after being freed from their Assyrian dominion in the 6th and 5th centuries BC. Following this, they wandered into ancient Europe, integrating with the purely Japhetic cultures already inhabiting the regions.

This is the second strewn field that broke free of NIBIRU during its 522 AD pass when Phoenix was also in the system. For 270 years the comet group wandered until passing Earth in *792 AD* (792 is NIBIRU's orbital period) and stabilized in a 555-year orbit around the Sun. The train appears to be 3 years long.

| | |
|---|---|
| 792-794 AD (4686-4688 AM) | 792 AD was the 24th year of Merovingian King Charles the Great, ruler over Franks, over all of France, Germany and regions of Spain, Italy and Austria. A Christian king descended from the tribe of Reuben. According to the *Anglo-Saxon Chronicle*, in 792 AD huge streaks of bright flames raced across the skies along with terrible comets (flying dragons) that brought famine. Later in the year in June, Norse Vikings (later to be integrated with the Normans) wrought destruction of churches and killed many Britons.(11)  The English scholar Albinus Alcuin and translator of the *Book of Jasher*, wrote about the Viking invasion. This year is calendrically significant, for as it is 792 Anno Domini, it is also 3031 years after the Great Flood in 2239 BC (1656 AM), and 3031 *BC* (historic countdown to Anno Domini system) was the year *864 Annus Mundi – 792 years later*, being the Deluge. |

(144 years over the ecliptic)

| | |
|---|---|
| 936-939 AD (4830-4833 AM) | The 3-year train enters the inner system. In 936 AD planet Phoenix passes over the ecliptic unseen from Earth (see *When the Sun Darkens*). |

(411 years under ecliptic completes 555 year orbit)

| | |
|---|---|
| 1347-1349 AD (5241-5243 AM) | The 3-year train begins entering inner system. This is exact year the Black Death plague reached Europe and began killing thousands a day as people turned a darkish purple and died miserably. The plague had already reduced the populations of China, India, probably all the Orient, the Middle East and northern Africa by a *third*, and would effect the same decimation in Europe. Though rats aboard ships did in fact aid in contaminating ports, thus spreading the disease, this plague had an *extraterrestrial* origin. The people of the period even attributed it to the passing of comets. We cannot ignore that these comets were once *oceans* on NIBIRU, a formerly inhabitable world (now a prison of the Anunnaki). These icy glaciers in space contain the refuse of frozen and decayed organisms and animals that once occupied those oceans. As we will see, much of the detritus falling to Earth is of carbonaceous and *organic* materials. The rains of a dead world kill another. The Plague was particularly devastating to France. |

| | |
|---|---|
| 1348 AD (5242 AM) | As the Reuben Comet Group passes through the inner system and over the ecliptic, a bright comet appears over Paris, France.(12)  Also, a pestilential wind infected the island of Cyprus with strange air and many died of asphyxiation. The anomalous nature of this plague is found in that it was attended with an earthquake.(13)  The Black Death by this time had killed 50% of the population of Europe.(14) |

(144 years orbiting over ecliptic)

| | |
|---|---|
| 1491-1493 AD (5385-5387 AM) | The 3-year train begins entering the inner system over the ecliptic. |
| 1492 AD (5386 AM) | At the instruction of King Ferdinand and Queen Isabella of Spain (partially descended from Reubenite stock), the Inquisitor-General Torquemada gave the Jews residing within the Spanish realm three months to either convert to Christianity or leave the country. Those remaining and refusing to convert would be killed, all their properties confiscated. On April 17th a Cristobel Colon, a Jew having had his name altered to Christopher Columbus, received permission and three ships full of provisions to search west over the Atlantic for a sea-route to the Spice Islands of Indian Ocean. The deadline for the expulsion of the Jews was August 2nd, the EXACT date Columbus was assigned to leave Spain, and from the *same* port that thousands of Jews boarded ships to leave the country (in Gulf of Cadiz). Why he was forced to wait until this day, knowing the port would be terribly busy, is a mystery not explained by historians.(15) Perhaps the secret of Cristobel Colon was actually detected by the crown. As Christopher Columbus sailed with the Nina, Pinta and the Santa Maria, over the Atlantic in alien waters, a *meteor* burst with a loud crash and fell across the sky, terrifying the sailors. This was merely a fragment, an asteroid turned meteor (or meteorite) of the Reuben group. |

(411 years orbiting under ecliptic)

| | |
|---|---|
| 1902-1904 AD (5796-5798 AM) | The 3-year train enters the inner system. This is the FINAL year of the Cursed Earth system of 414-year epochs spanning back to the destruction of the Pre-Adamic World in 4309 BC (see *When the Sun Darkens*). Planet Phoenix crosses the ecliptic unseen from Earth but immerses our planet in a canopy of hundreds of millions of tons of dust, reported to have fallen all around the world in 1902-1903. A comet from Phoenix is seen and named Morehouse. |
| 1903 AD (5797 AM) | A strange unidentified lavender-colored substance rained upon Ouden, France, on December 19th.(16) Of the dusts falling upon the earth from space this year many scientific tests concluded that the material was from 9-36% *organic*.(17) As these debris fields travel orbiting the Sun, they appear as giant nebulous clouds in space. The Reuben group may have been spotted in 1903 AD by the Lowell Observatory when it reported that on May 20th it had viewed through its powerful telescope a body described later as a cloud, near Mars.(18) |

| | |
|---|---|
| 1904 AD<br>(5798 AM) | A meteorite crashed into Alta, Norway, its surviving fragment weighing 198 lbs. This fact alone hints that we are being warned from heaven (space) that a great Stone will fall in *198 years*, or 2106 AD, the year 6000 Annus Mundi. The Normans, of French ancestry who were themselves descended from Reubenite and Benjamite stock, occupied Norway and today constitute many of their own descendants. In this year, America's 4 billion chestnut trees are virtually wiped out by blight. The fortuitous discovery in 2006 AD of four living chestnut trees that survived this has baffled botanists.(19) |

(144 years orbiting over ecliptic)

| | |
|---|---|
| 2046-2048 AD<br>(5940-5942 AM) | The 3-year train enters the inner system. As the United States is destroyed in 2046 AD (see later in this work), a mixed population containing a huge number descended from France and Norway (thus Reubenite), the comet group maintains its Reubenite theme. |
| 2047 AD<br>(5941 AM) | Over a million displaced Americans flood Europe and Israel (see *Chronicon*) and in this year, France and portions of Europe suffer strange atmospheric fallout, possibly meteorites or even a comet that enters the atmosphere and detonates, as in 1908 AD over Tunguska, Siberia. |

Though in 1919 AD and the decades before, when the antiquarian and critic of the scientific establishment, Charles Fort, was compiling his data of unusual phenomena, he did not have access to the truly ancient records we have found as of the 21st Century, nonetheless, Mr. Fort had reviewed enough material and reports from the past few centuries to conclude in his epic work, *Book of the Damned*, that ". . .four classes of phenomena that have preceded or accompanied earthquakes are: unusual clouds, darkness profound, luminous appearances in the sky, and falls of substances and objects called meteoric or not."(20)

## Accurately Interpreting the Mayan Long-Count and its Relations to the Great Pyramid

We haven't dug deeply enough or studied things thoroughly enough to reach reliable conclusions about the distant past. So many of the facts commonly accepted today are only educated guesses.

—John Keel, *Our Haunted Planet* (1)

There are in print over a thousand books explaining how the Mayan Long-Count ends in the year 2012 AD, and the subject fills tens of thousands of articles and has deluged the internet. We have been led to believe that some epic, transforming event will occur in this fateful year and will change the world and humanity. A New Consciousness will emerge, the founding of some abstract Cosmic Evolution event. . . all fictions perpetuated by highly imaginative individuals who fell prey to the misinformation that the Mayan Calendar ends in 2012 AD. This is an absolute mathematical *impossibility*. The year 2013 will follow, as will 2014 and so on.

The Maya preserved an advanced calendrical system from remote antiquity that cannot rightly be called their own. They *inherited* it. The system has been proven now to come from the much earlier Olmeca civilization of the Yucatan that antedates the founding of the Mayan culture, which admittedly, may in turn find their ancestors in the Olmeca. The discovery of a jade figurine with a calendar inscribed upon its stomach indicating to scholars the date of 3113 BC demonstrates that this unique calendrical system precedes the Maya and every other archaic known American culture after 1200 BC. Jeanne Reinhert in *Science Digest* (Sept. 1967) wrote that "It is a masterpiece of mathematical knowledge, and it was the Olmecs, not the Maya, who developed it."(2) But this too is an overstatement, for the Olmeca too were merely the inheritors of this truly remarkable system.

The origin of the Mayan Long-Count calendar, according to scholars, was in the year we know as 3113 BC, or 782 Annus Mundi. They are absolutely correct. This dating system began exactly *13 years* before 3100 BC, the date assigned by historians as the start-date for the ancient Brahmanic calendar and that of King Narmer (Menes) of the beginning of Egyptian history who unified the people into a United Kingdom.(3) Narmer is in fact the biblical patriarchal prophet-king Enoch, described with the epithet NAR (NER being the Sumerian title for Anunnaki: the kingship) and MER, the Egyptian designation for a *pyramid*. Zechariah Sitchin in his work, *Journey to the Mythical Past*,(4) clearly demonstrates that upon the Narmer Palette, also known as the Victory Tablet of King Menes, is depicted a clearly obvious *pyramid* with smooth sides.

Enoch was the greatest Sethite king before the Flood, even ruling over the Anunnaki. His reign lasted for 300 years and, as revealed in my book *The Lost Scriptures of Giza*, Enoch vanished, leaving behind divine instructions that the Sethites followed, which resulted in the construction of the Great

Pyramid complex. It took 90 years to complete and was finished in the year 1080 Annus Mundi, or 2815 BC. Enoch personally oversaw the building of Stonehenge I and II, and in 3100 BC, when the ancient Indian (Vedic) and Egyptian chronologies began, Enoch was enjoying his *108th year* of kingship. See *Chronicon.*

The importance of this year is exemplified in the Egyptian *Old Chronicle* that cites 113 regnal descents in a 36,525-year period from Pharaoh Scorpion to 290 BC, when Manetho studied this text from older sources. The scorpion is an *arachnid*, the symbol for the Anunnaki before the Flood as exhibited in this author's work, *Descent of the Seven Kings*. This impossible sum of years, 36,525, is to be decoded by the *lunar* months of the year, 13, to discover the actual sum of years actually being referred to in the encoded text. The 36,525 "years" divided by 13 is *2810 years*, which from 290 BC spans back to our year of 3100 BC, the start of Egyptian dynastic history.(5) The calendrical import of this encoded system serves to allude to the year of 365.25 days as known in the time of Manetho in 290 BC, a year *unknown* to the ancient Maya.

As seen prior in this work, and detailed very specifically in this author's other books, the year was 360 days evenly matched with the Zodiacal stellar year, an annual orbit of the earth geometrically perfect. The introduction of the 365.25-day year was due to cataclysm in 713 BC when the Dark Satellite nearly collided into Earth. Enoch erected Stonehenge to commemorate not only the original heavens under the 360-day system, but also to show that in the future this system would collapse into a *different* type of annual calendar. The precise science of Enoch will later be demonstrated in this book. For now let it be understood that the calendrical system the Olmec and Maya preserved was ultimately from *Enoch*, and it was specifically designed to start *13 years* before the other ancient dating systems that had their genesis in 3100 BC.

Enoch was a prophet in the dawn of human civilization whose ministry was designed for the Last Days. In the 160th year of the *life* of Enoch, 3113 BC (3100 BC was the 108th year of his *reign*), he designed Stonehenge II, adding the mysterious bluestone ring and inner horseshoe, stones symbolic of the Anunnaki. These 79 enigmatic bluestones added to the Stonehenge I site, which originally consisted of exactly 81 sarcen stones, added up to make Stonehenge II a complex concentric ring-site of *160 stones*, reflecting the year of the life of its principle architect. The Mayan Calendar's chief function was *not* to record the passage of time, but to count down to a time known as 13.0.0.0.0 at the conclusion of 13 Baktuns (144,000 days each) when TIME WOULD COLLAPSE. The end of the Age would be known by a fundamental alteration in the passage of time itself. So we see that Enoch's ministry for the Last Days is reflected in the conceptual purpose of the Mayan Long-Count.

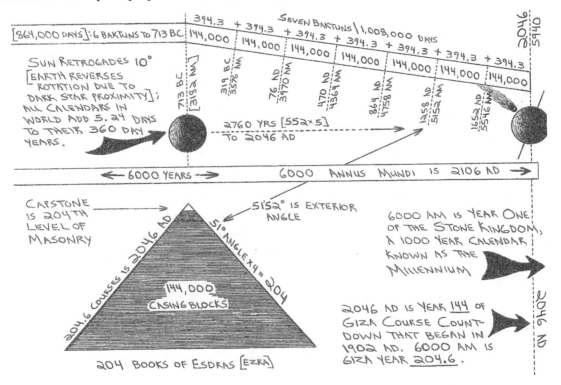

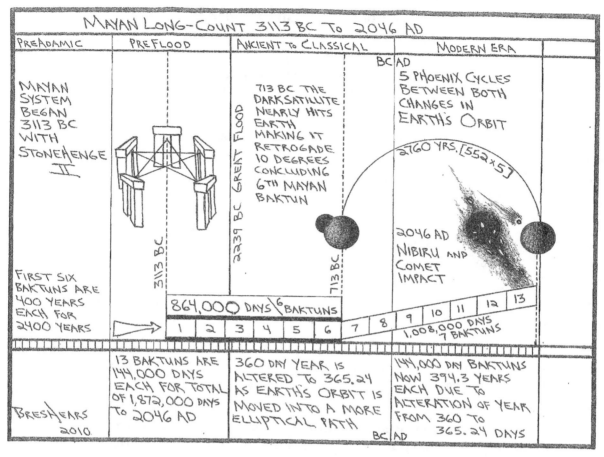

Because modern scholarship ignores solid archeological and textual evidence that Earth's orbit was dynamically altered in 713 BC when all calendrical systems from China to the ancient Americas adopted a new *vague year* of 365.25 days by adding these "evil days" of 5.25 to the end of their annual counts, the modern interpretation of the Mayan Long-Count system is totally corrupted beyond anything the system meant to convey. The Dark Satellite almost collided into the planet in 713 BC and the *effects* of this on our world are fully chronicled by such prolific researchers as Immanuel Velikovsky and Zechariah Sitchin. As a result, scholars have taken the start of the Mayan Long-Count at 3113 BC and calculated the entire system based on the modern year of 365.25 days, which abbreviates the 13 Baktuns of 1,872,000 days equaling 5200 years (on the 360-day system) to a period of *5125 years*. . . a number absolutely *unknown* to the Maya. This interpretation is arbitrary, for it even breaks up the Mayan mathematics incorporated into the system. A baktun is exactly 144,000 days, or 400 *tuns*, and a tun was equal to *360 days*. The number that links all the early American cultures was 52, and 5200 actual years of 360 days each is the entire longevity of the Mayan Long-Count, or 1,872,000 days. The system is totally incompatible with the scholarly accepted end-date at 2012 AD, so blindly followed by the literati. The tun was measured out as a Mayan year of 18 months, consisting of 20 days each, for 360 days. Five "unlucky days" were added later.(6) The 260-day period was particularly sacred to them as well, a unit measured as 52 x 5. Zechariah Sitchin noted that the three methods of calculating the Long-Count of the Maya all relate to the number 360. Despite this revelation, he along with the entire world still calculates the system using the modern vague year of 365.25 days.(7)

Thus far, this author has committed the error of all those before him in explaining calendars and concepts from antiquity. But to redeem and separate himself from those who dare not demonstrate their fallacies, a detail-specific chronology is now exhibited on the Mayan Calendar.

## Mayan Long-Count Calendar

3113 BC
(782 AM)

Mayan Long-Count begins, counting down, 1,872,000 days, or 144,000 x 13 days. This will conclude with 13.0.0.0.0 when *time will collapse.*

2713 BC
(1182 AM)

First baktun complete (144,000 days, being 400 years on 360-day system).

2313 BC
(1582 AM)

Second baktun complete (288,000 days, being 800 years on 360-day system).

1913 BC
(1982 AM)

Third baktun complete (432,000 days, being 1200 years on 360-day system).

1513 BC
(2382 AM)

Fourth baktun complete (576,000 days, being 1600 years on 360-day system).

1113 BC
(2782 AM)

Fifth baktun complete (720,000 days, being 2000 years on 360-day system).

713 BC
(3182 AM)

Sixth baktun complete (864,000 days, being 2400 years of 360-days each). This is 2760 years, or 552 x 5 to *next* time orbit is altered in *2046 AD.* These 864,000 days refers to *864,* making this year a *new foundation for time.* The lost moon of NIBIRU, called the *Dark Satellite*, nearly collides into Earth and pushes our planet about .01% away further from the Sun, increasing its elliptical orbit and adding *5.25 days* to the 360-day year. The planet stops rotating and then initiates a 10-degree retrograde motion briefly, before resuming its normal rotation. Review the 713 BC event in the Archive detailing the Dark Satellite. This concludes exactly *2400 years* of the Mayan system, leaving 1,008,000 days remaining in the Long-Count (7 baktuns).

319 BC
(3576 AM)

Seventh baktun complete (1,008,000 days, being 395.4 years of 365.25 days each). The Dark Satellite passes through inner system unseen from Earth, 144,000 days after 713 BC.

76 AD
(3970 AM)

Eighth baktun complete (1,152,000 days, being 395.4 years of 365.25 days each). 144,000 days after 319 BC. The Dark Satellite appears and it is witnessed in the sky as a *javelin*, which Titus recorded and Pliny wrote was an omen of doom.(8)  This is also the beginning of the *Saka Calendar* of India, which counts down *476 years* to the invasions of the Huns into India in 552 AD (4446 AM). The Trojans fell to the Mycenaeans and Greeks in 1229 BC (see *Chronicon*), which began a *476-year* countdown to the founding of Rome in 753 BC by the descendants of Troy, the Romans. In the year *476 AD* the Romans were defeated by the Germans, initiating the famous Fall of Rome. As 120 is the *foundation* number (not to be confused with the Foundation of Time number 864), we see how intriguing it is that 76 AD was the year *120* of the Roman Calendar, known as the Julian. The Fall of Rome itself in 476 AD merely began a *1300 year* period to the 1776 AD founding of the United States of America when the *13 Colonies* rebelled against England, a people of Anglo-Saxon heritage largely descended from early European *German* tribes.

470 AD
(4364 AM)

Ninth baktun complete (1,296,000 days, being 395.4 years of 365.25 days each) 144,000 days after 76 AD. Dark Satellite fallen out of synch with Maya system.

864 AD
(4758 AM)

Tenth baktun complete (1,440,000 days, being 395.4 years of 365.25 days each). 144,000 days after 470 AD. It is beyond coincidence that the 10th baktun is completed (mystic association to number 10 is *completion*) in the year *864* Anno Domini. In this year the *Irish Annals* record both a lunar and solar eclipse, the solar eclipse no doubt being the transiting occult of the Sun by a *comet*.(9)  This year marks the start of Earth's final three Cursed Earth periods of 414 years each, or 1242 years to Armageddon and the year 6000 Annua Mundi (2106 AD). See *When the Sun Darkens* for a full review of the Cursed Earth system. This also marks the start of the final three Mayan baktuns until the *collapse of time.*

1258 AD
(5152 AM)

Eleventh baktun complete (1,584,000 days, being 395.4 years of 365.25 each). 144,000 days after 864 AD. This year of 5152 Annus Mundi mirrors the geometrical angle of the Great Pyramid's four facing slopes measured at *51.52* degrees, the monument built at the instruction of Enoch and covered with *144,000 white limestone facing blocks* of expert precision. In this year the Mongol armies under the Khan invaded Iraq (Babylonian) and crushed the Abbasids.

1652 AD
(5546 AM)

Twelfth baktun complete (1,728,000 days, being 395.4 years of 365.25 days each). 144,000 days after 1258 AD. Almost as if a divine warning was provided to the descendants of Rome and Europe, *America*, in this year of 1652 AD a bright meteorite crashed between Siena and Rome in May.(10)  This begins a 144,000-day countdown to 2046 AD.

2046 AD
(5940 AM)

*Thirteenth* baktun complete (1,872,000 days, being 395.4 years of 365.25 days each). 144,000 days after 1652 AD. The Mayan Calendar is complete. But will *time* itself collapse? The answer to this penetrating question will be answered in the affirmative by the truly incredible information contained in the next sections. The Mayan Long-Count was divinely inspired to both conceal and then later convey a Divine Act that will transpire in the Last Days, an event that will bring into perfect clarity an accurate interpretation of the enigmatic Book of Revelation. As the Long-Count of the Maya and earlier Olmeca was preserved as an *American* system, we will also see how its end in 2046 AD specifically affects the United States of America, 2760 (552 x 5) years after *713 BC*.

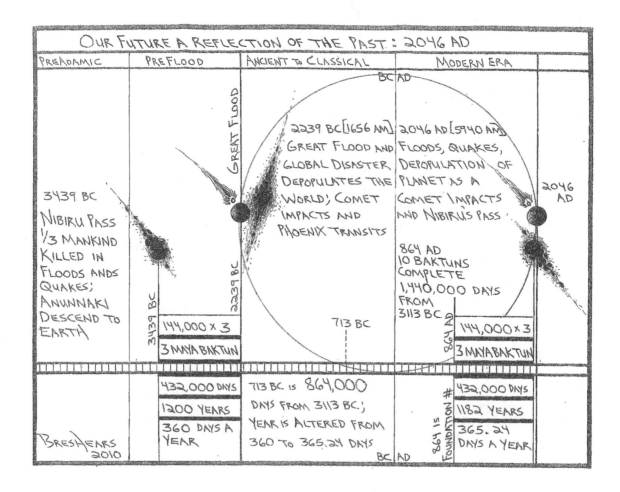

The casing blocks that adorned the Great Pyramid covered a surface area of 22 acres. These hard white limestone blocks protected the softer limestone blocks within the structure constructed of 203 levels of stones, course upon course, to the height of a 41-story building. These casing blocks were massive, about 100 inches thick, several tons each ". . .and of complex shape, smoothly finished and formerly held in a very thin layer of cement of great strength and unknown formula."(11) What baffled archeologists the most when these casing blocks were discovered in 1837 AD and in subsequent studies was the fact that the builders of the monument employed a 0.010-inch precision on the casing block planes, utterly *perfect*, when a 0.25-inch plane would have been sufficient tolerance, as is used today in modern brickwork.(12) The dimensions of the surviving casing blocks reveal that the four faces of the structure had a total of 144,000 protective white limestone blocks.

In David Davidson's epic work entitled *The Great Pyramid: Its Divine Message*, a colossal book published in 1924 AD, we find that the architect of this monument encoded a message into its construction interpreted through a study of its four foundational cornerstones. Buried in this geometrical prophecy was the exact date of the longitude of perihelion of an unspecified object that astronomically related to the Earth. This future date, 2045 AD, marked the Great Pyramid's *chief calendrical prophecy* encoded within the dimensions of the structure's four cornerstones, the architect's metaphorical Four Corners of the World. His book is 568 pages with hundreds of charts and illustrations but *nowhere* in his writings does he allude to what this object is that approaches our Sun in 2045 AD, or how it will affect our planet.

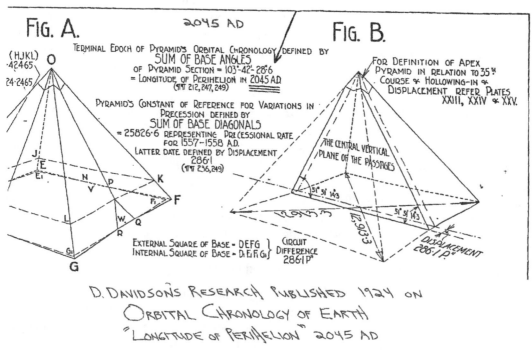

Davidson's amazing discovery holds that this Astronomical Chronology began in the year 4040 BC, with Davidson citing Dr. Crommelin of the Greenwich Observatory. Why the variance of one year from 2046 AD is explained by this begin-date of 4040 BC, which is itself one year off the true year. As this author has shown in *Descent of the Seven Kings*, he has demonstrated in several charts and timelines how Earth *began a new orbit* in the year 4039 BC. The evidence of this is overwhelming. This date was exactly 270 years after the Pre-Adamic World was completely destroyed (see *When the Sun Darkens*) in 4309 BC. What makes this 4039 BC new orbital chronology of Earth so fascinating is that it places the end of the Great Pyramid's Orbital Chronology at *2046 AD*, which is itself verified by the fact that this Astronomical Chronology of our planet *begins and ends* in a 144-year period.

In this author's prior work, *When the Sun Darkens*, it is shown how the Cursed Earth Calendar of 414-year periods began in 4309 BC, when the Pre-Adamic World was ruined, and ended in the year 1902 AD (5796 AM). The Earth was lost in space in a chaotic state for 270 years before its orbit stabilized in 4039 BC (Great Pyramid's Astronomical Chronology begin-date). The *144th year* of this Renovation of Earth from a stabilized orbit was *Year One* of the Annus Mundi Calendar and Original Hebraic system, 3895 BC. An earthly Paradise was enjoyed for 144 years until the beginning of Cursed Time in 3895 BC, a subject also explained in *Descent of the Seven Kings*. The amazing synchronicity of this system is that the end of the Cursed Earth chronology was 1902 AD, which began a *144-year* countdown to 2046 AD. As 1902 AD began the Giza Course Calendar, a timeline for the Last Days recorded in *levels* of stones of the Great Pyramid to the year 2106 AD, the descent of the Chief

Cornerstone from heaven in the year 6000 Annus Mundi, 2046 AD, is represented by the unusually thick 144th course of masonry in the monument. The four cornerstones of the pyramid that define the Astronomical Chronology which indeed ends in 2046 AD themselves provide a *foundation* to the edifice, the four equidistant stones forming a square.

The antiquarian and one-of-a-kind researcher William Corliss in his work *Ancient Structures* wrote that ". . .the Great Pyramid is a scale model of the earth's Northern Hemisphere. In this model the pyramid's *height* represents the polar radius; its perimeter, the earth's equator."(13) The Northern Hemisphere will be most affected by the events of 2046 AD. None of these authors knew anything of NIBIRU or the Anunnaki, nor of the comets and asteroids destined to end their orbital cycles by colliding into the Northern Hemisphere. The monument seems to acknowledge the fact that throughout the history of the world during this Astronomical Chronology, *all* of the world's greatest nations and empires were seated on the equator or above it in the Northern Hemisphere. Most of Egypt is north of the Equator. From China to Central America and even some of South America, all the regions between are all on and north of the Equator.

The structure's foundation begins the Giza Course Calendar, a system counting down to the return of the Chief Cornerstone to vanquish the Anunnaki in 2106 AD (6000 AM), a system fully chronicled year-by-year and even through the future in this author's *Chronicon*. This base foundation, indicated by the year 1902 AD, which was the subject of *When the Sun Darkens*, was itself exactly *4140 years* after the Great Flood when the planet was virtually destroyed in 2239 BC (1656 AM). This was a period of 10 Cursed Earth epochs. And with this fact we discover yet another vertical geometrical message hidden within the masonry of the Great Pyramid. As the Flood was caused by the *impact of a comet* from NIBIRU (see *When the Sun Darkens*), the next major impact of a comet that will destroy a significant portion of our world will be in 2046 AD, the *144th level* of bricks in the structure which happens to be *4140 Pyramid Inches* above the baseline of the monument (1902 AD).

Perhaps one would like to criticize the use of Pyramid Inches, and, considering that we have not in this study explored this fascinating mode of measurement, this is certainly permissible. But for those seeking the ancient evidence of a lost corpus of prophetic knowledge buried within the geometry of the Great Pyramid concerning this epic year of 2046 AD, read *Chronotecture: Lost Science of Prophetic Engineering*.

As the Mayan system was founded upon a framework of periods enduring 144,000 days, which were changed to conceal a future date (2046 AD) when time would be significantly altered, so too was the Great Pyramid covered in 144,000 casing blocks that were eventually stripped off the building after an earthquake in 1356 AD. This allowed researchers to explore the monument's deeper geometrical mysteries. The apocryphal writings claim that in the Last Days a special ministry of 144,000 witnesses will spread the Word, and this is also mentioned in the Revelation. These are the 144,000 people in the End Time according to the Maya who "meet the Sun," opening the doors of salvation.(14) This is the meaning behind the 144,000 casing stones on the structure, which is further demonstrated in Psalm *144*. ". . .our sons may be as plants (trees of life) grown up in their youth; that our daughters may be as *cornerstones*, polished after the manner of a palace. . ."(15) This passage is in Psalm 144, verse 12, which is the square root of 144.

# Arrival of NIBIRU, Cataclysm and Anunnaki Chronology of 2046 AD

> . . .the Sun turns back, earth sinks into the sea, the hot stars fall from the
> sky, and fire leaps high about heaven itself.
> —Viking prophecy of Elder Edda  (1)

The Anunnaki Homeworld, NIBIRU, is a gigantic planet that traverses the nether regions of the solar system in a 792-year orbit around our Sun to the Dark Star and back. It is a dead world, encapsulated in contaminated frozen glacial sheets of water from primordial oceans. As our own planet was literally quick-frozen in 4309 BC with the destruction of the twin-luminary system into one now having a darkened compressed star, so too did the planet of the Anunnaki suffer the same fate. But unlike terrestrial organisms here on Earth, the Anunnaki are *immortal*. They have physical bodies and, like amphibians and reptiles, they can, through inclement climatological conditions, remain dormant and in a type of temporal stasis.

This orbiting mortuary of imprisoned demons has, trapped within its immense miles-high glacial sheets, frozen flora and fauna, ruined architecture and artifacts from those cities that previously thrived upon the planet before 4309 BC, entire buried land surfaces as well as traces of *blood*. Gigantic fragments break away and lose mass at perihelion and ultimately enter into their own orbits around the Sun, depositing these ice-fossils, traces of blood and even radioactive contaminants into Earth's atmosphere. And as we have clearly seen. . . sometimes depositing the Anunnaki themselves.

The Anunnaki were used by the Godhead to reap a harvest of chaos and destruction among rebellious men as in 1899 BC in Babylonia to stop the Tower project, 1447 BC during the Exodus against the Egyptians and again in 713 BC when, through the agency of angels, God destroyed 185,000 soldiers. These were historical events serving as prophetic foreshadowing of what would eventually transpire in 2046 AD . . . a full scale Anunnaki *invasion*.

The synchronicity of this 2046 AD date is profound. In this year NIBIRU *ascends* over the ecliptic and will be seen from the southern hemisphere first, *exactly* in the way as predicted by our modern prophet Zechariah Sitchin.(2)  The Anunnaki planet was last seen in 1314 AD as it occulted the Sun over Asia and blackened the heavens in shadow as recorded by European historians, amidst the chaos of quakes, volcanic disturbances, astronomical omens, plague mists and general epidemic. Debris and fallout from *several* comet groups having their origin with NIBIRU will rain their ruin on Earth in 2046 AD and this is the end-date for the Great Pyramid's Astronomical Chronology, which began when Earth initiated its present orbit around this Sun in 4039 BC. This ends the Mayan Long-Count calendar's 13 baktuns of 144,000 days each since 3113 BC and 2046 AD is precisely *52 years* (ancient *divine* number of Americas) after comet Shoemaker Levy 9 was struck with a flux tube and fragmented prior to crashing into Jupiter in 1994.

2046 AD is *666 years* after 1380 AD when the great Salisbury Clock was erected in England, the oldest surviving (still works today) mechanical clock in the world.(3) Its location at Salisbury is near the most ancient megalithic clock in the world, *Stonehenge*. As 2046 AD marks the alteration of time itself (as we will see), the 666 years seems to be a deliberate countdown to this change as well as the arrival of the Anunnaki. Also in 1380 AD the Zeno Brothers illustrated a map of an *ice-free* Greenland.(4) This map includes also some polar regions. The map itself could not have been originally made at this date, as there was ice all over North America and the polar areas detailed. It was merely a 1380 AD recomposition. The parallel here denotes that America will be affected at the 666-year end of this countdown from 1380 AD. A fact we have seen already by the impact of comets and asteroids.

The year 2046 AD is further 270 years after the founding of the United States of America from the original 13 Colonies of England. These 270 years parallels the 270 years from 4309 BC destruction of the Pre-Adamic World to the 4039 BC Renovation of Ruined Earth when the Anunnaki appeared in the capacity of Builders who *renovated* the earth but rejected the Plan of the Chief Cornerstone, known as the Word of God. This is the Stone the Builders rejected who will return in 2106 AD (6000 AM) to defeat and imprison these rebellious Anunnaki overlords. During these 270 years the Earth was in ruin, lost in space moving away from the Dark Star. *Descent of the Seven Kings* fully expounds upon this history. It was during this 270 years that, according to the Babylonian *Enuma Elish* tablet texts, ". . .long were the days, *years were added*," this terminology relating that time was even then measured in *days*, but due to the removal of Earth from its position, the Sun *disappeared* (Earth left the Daystar which had collapsed into a Dark Star) and traveled along with NIBIRU to this present Sun.

This *Enuma Elish* text was translated into English and widely published in 1902 AD, the end of the Cursed Earth system of 414 year periods, detailed fully in *When the Sun Darkens*, the year 1902 being *144 years* before 2046 AD. In the *Enuma Elish* tablets are given detailed descriptions of the Anunnaki and we find, amazingly, that they mirror the physical descriptions given to them again in the Book of Revelation, which was composed by the last apostle, John, who was imprisoned on the isle of Patmos in 96 AD when NIBIRU was at aphelion in that exact year (furthest distance from the sun). Additionally, by John's time in the first century AD cuneiform was *extinct*, the Babylonian tablets of the *Enuma Elish* had already by his day been buried in the deserts of the Middle East. And counting 666 years into the future from when John received the Revelation vision we find the Abbasid Dynasty in 762 AD making its capitol at Baghdad, the New Babylon, 144 months after it assumed authority after the fall of the Umayyids. Incredibly, 762 AD was the 6000th year of the Anunnaki Chronology, which began in 5239 BC (See *Descent of the Seven Kings* or *Chronicon*).

GREAT PYRAMID'S GEOMETRICAL ORBITAL CHRONOLOGY 2046 AD

The cryptic and esoteric number 108 encoded in the Great Pyramid's chronometrical timelines, within the geometry of Stonehenge and, as we will see, specifically encoded within numerous crop circles, leads us to our most interesting countdown. Exactly 108 years prior to 2046 AD was the year 1938 AD when on October 30th, Orson Welles dramatized H.G. Wells' *War of the Worlds* story for his Mercury Theater Radio program. The broadcast concerned seemingly live accounts from around the United States and the world concerning an *invasion of extraterrestrials* from Mars, a broadcast that according to a later Princeton study terrified six million people, with 1.7 million people actually believing that Earth was being invaded and 1.2 million people actually took action because of it.(5) Mars was broadcast as the invading planet. The parallel here is that in 1976 (the same year Sitchin's first book, the *12th Planet*, was released) NASA's Viking I and II set down upon Mars, the former having a probe that from *1080 miles* (108 x 10) above the surface of the Martian landscape took pictures of apparent gigantic architectural ruins in a post-cataclysmic topography at the Cydonia region. Satellite photo 70A13 revealed a distinct Sphinx-like face, as well as a pentahedral pyramid known as the D & M Pyramid. This Red Planet might very well have long ago been Anunnaki populated, controlled or colonized. The year 1938 AD, 108 years before 2046 AD, was also notable for the amazing discovery by a South African fishing trawler of a living coelacanth, a marine prehistoric fish used up until this time as an Index Fossil to date other supposedly 70 million-year-old creatures.(6)

Sitchin's NIBIRU revelations and publication of his research on the Anunnaki as well as all these other events in 1976 AD began a 70-year countdown to 2046 AD. This was also the United States' Bicentennial Celebration, revealing that it will be in the USA's *270th year* that NIBIRU returns and the Anunnaki invade (1776 + 270 is 2046 AD). In 2046 AD a series of meteoric impacts, cometary collisions and the devastating effect NIBIRU will have on our planet will effectively *end* the United States, Canada, Mexico, much of Central and South America, with these regions being the 13th Geographical Sections of Earth. The Ancient World from the Far East of Japan and China to the western shores of Europe is made up of 12 divisions, the earliest planisphere conceptualized by the ancients. America is the most recently colonized continental area, the most powerful Empire being the United States, and it will be *removed*. The USA fulfills the prophecies and types of the 13th Tribe of Israel, the *Tribe of Adoption* that had its origin with Joseph in Egypt, this 13th Tribe having begun in the Last Days as 13 Colonies that broke away from their brother tribe Manasseh, known as Great Britain. These 13 were preceded by the 13 Clans of the Maya who lived under their 13 Heavens, knowing that one day in the future after the end of the 13 baktuns of the sacred calendar that *time would collapse* and END THEIR PERIOD OF OCCUPATION in lands, the Maya admitted, that were *not their own*. With the removal of America by total catastrophe will be initiated the *13th Chapter of Revelation*, as the Antichrist gains absolute authority over the Earth. Thirteen was the number of Enoch, the sum of years the Mayan Calendar began in 3113 BC before the start of the other Ancient World calendars in 3100 BC and is also encoded in the Annus Mundi system as the fateful year of America's end.

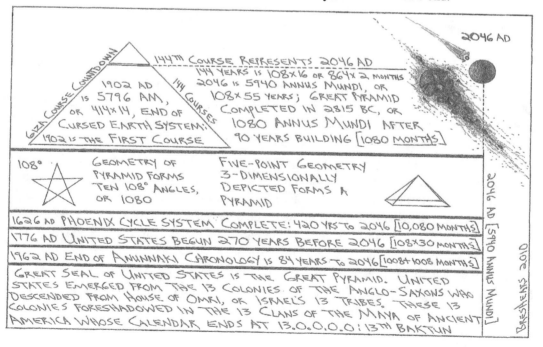

The longevity of the United States appears to have been encoded by the very men that signed the U.S. Constitution. In the east the number for a man's life was 108, as the beads of a Buddhist rosary. It was an early belief that a holy man would live to see 12 x 9 years, or 108. The men that signed the Constitution numbered 55, and these 55 men all represented conceptually the sum of 108 to provide mathematically the number 5940 (108 x 55), the *exact year* of the Annus Mundi timeline equaling our *2046 AD*, the 270th and final year of the United States and North America for that matter.

In the Book of Revelation the United States, by 2046 AD, has become a modern-day *Sodom*. By this date the people of God have *come out of Sodom* (migrated away from the USA) in preparation for the escape of America's impending ruin and in fulfillment of the Old Testament prophecies concerning the return of the lost tribe of Ephraim *back to Israel*, foretold by Ezekiel the prophet.(7)  Sodom is the modern symbol for a people who have grown lazy, absolutely rich but greedy, spiritually vacuous, laden with resources and luxuries, sexually deviant and completely alienated from God. The ancient Cities of the Plain were Sodom, Gomorrah, Admah, Zeboiim and Bela, and four of these cities were laid waste in the year 2047 Annus Mundi (1848 BC), 2047 in numerology equaling *13*. In the timeline of the First Adam, the Annus Mundi system that started in 3895 BC with the 6000-year countdown to 2106 AD (6000 AM) when the *Second Adam* (Christ) returns, Sodom and Gomorrah were destroyed in the 2047th year. But in the timeline of the Second Adam, the start being the birth of Christ on the cusp of 2-1 BC, the destruction of the United States in 2046 AD is actually the 2047th year of the Second Adam (title for Christ in the New Testament). Amazingly, the *only* city not destroyed of the Cities of the Plain was Bela, a city of *immigrants* not native to Canaan, but from Mesopotamia.

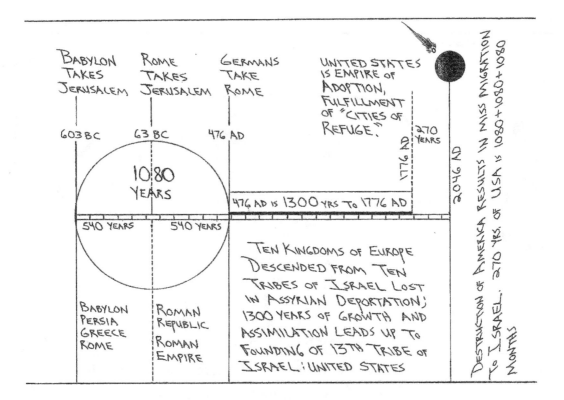

The migration of Americans back to Israel prior to the cataclysm, as well as millions of Europeans, will bring together the descendants of the 13 Tribes of Israel for the greater period of the Apocalypse. Though not covered in this book, 2046 AD is paralleled by its Isometric Projective date of 1950 AD. As fully explained in *Chronicon: Timelines of the Ancient Future*, there are specific year-dates in world history that serve to exhibit that the past is a predicate for the future, that the unfolding of events in the future are based off patterns initiated in the past. Isometric years are like ripples in the pond of time, each wave-ring before the epicentral marker (key year) is exactly the same distance in time as the opposite wave-ring. The Last Days Epicentral Year is 1998 AD, exactly *108 years* before Armageddon and the return of NIBIRU from over the Sun in 2106 AD (6000 AM). 2046 AD is 48 years after 1998 AD, and 48 years *before* 1998 was the year 1950 AD, when Israel enacted the *Law of Return*, calling for all Jews around the world to migrate back to Israel. Judah was last to be dispersed among the nations, by the Romans in 70 AD, and then finally in 135 AD, effectively destroying the land of Judea. The Lost Tribes of Israel were first to be deported and then assimilated into Asian and European peoples. In reverse order, Judah is restored to the Land of promise *first* and then the Lost Tribes will follow, preceding the 2046 AD appearance of comets and NIBIRU.

The destruction of the USA is found in a vision of Ezra who witnessed the fiery ruin of a nation symbolized in the End Times as an Eagle, a bird Ezra is specifically told by an angel to be the same Eagle as represented in Daniel's vision.(8)  The prophet Daniel saw Four Empires that would *scatter* Israel around the world (see box below): Babylon, Persia, Greece and Rome (Iron Empire), these symbolized as a Lion with Eagle's wings, a Bear, Four-Headed Leopard and a Dreadful Beast. In antiquity the Eagle's Wings referred to those tribes of Israel already occupying Babylonia from the earlier Assyrian deportations of 745 BC (3150 AM), which began the Post-Exilic Chronology of *2520 years* (360 x 7) to the founding of the 13th Tribe, Ephraim, as the United States in 1776 AD (5670 AM).

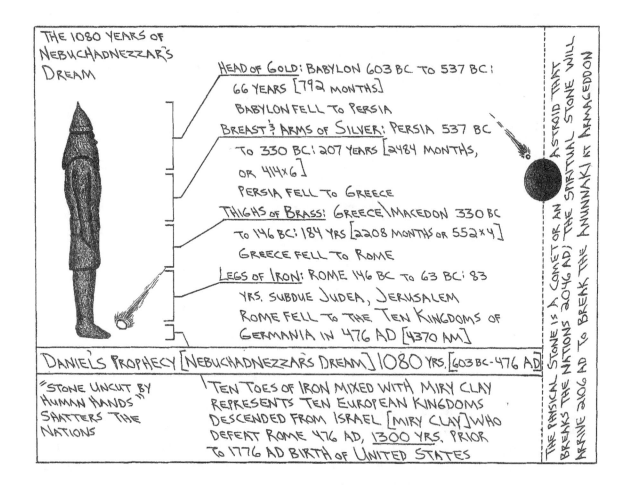

THE 1080 YEARS OF NEBUCHADNEZZAR'S DREAM

HEAD OF GOLD: BABYLON 603 BC TO 537 BC: 66 YEARS [792 MONTHS]
BABYLON FELL TO PERSIA

BREAST & ARMS OF SILVER: PERSIA 537 BC TO 330 BC: 207 YEARS [2484 MONTHS, OR 414×6]
PERSIA FELL TO GREECE

THIGHS OF BRASS: GREECE\MACEDON 330 BC TO 146 BC: 184 YRS [2208 MONTHS OR 552×4]
GREECE FELL TO ROME

LEGS OF IRON: ROME 146 BC TO 63 BC: 83 YRS. SUBDUE JUDEA, JERUSALEM
ROME FELL TO THE TEN KINGDOMS OF GERMANIA IN 476 AD [4370 AM]

DANIEL'S PROPHECY [NEBUCHADNEZZAR'S DREAM] 1080 YRS. [603 BC-476 AD]

"STONE UNCUT BY HUMAN HANDS" SHATTERS THE NATIONS

TEN TOES OF IRON MIXED WITH MIRY CLAY REPRESENTS TEN EUROPEAN KINGDOMS DESCENDED FROM ISRAEL [MIRY CLAY] WHO DEFEAT ROME 476 AD, 1300 YRS. PRIOR TO 1776 AD BIRTH OF UNITED STATES

THE PHYSICAL STONE IS A COMET OR AN ASTROID THAT BREAKS THE NATIONS 2046 AD; THE SPIRITUAL STONE WILL ARRIVE 2106 AD TO BREAK THE ANUNNAKI AT ARMAGEDDON

But this vision had a twofold meaning – one for those empires that would scatter Israel and again, for four future empires that would *bring the tribes back together* in the Last Days. Here the Lion with Eagle's Wings was the British Empire, that through military intervention and Parliamentary acts established a permanent Jewish presence and nation, Israel, back in Palestine. The Eagle's wings disappeared from the lion in the vision, which was the establishment 2520 years later of the Lost Tribes, becoming the most powerful empire this world has ever known, the United States, Land of the Eagle, having come from British colonies. The Great Bear is now fulfilled in Russia and the pogroms instigated the migrations of millions of Jews into Europe, America – and even today, Israel is literally filled with Russian Jews. This was followed by the Leopard of Germany, with the Nazis compelling mass migrations of Jews before they began their concentration camp program and then after the end of the War. The final empire, America, is fulfilling the role of Iron Empire, having descended from Rome politically, but culturally is fulfilling the prophecies of Ephraim, the Empire of Adoption, born in Egypt (outside Israel), but destined to *return*.

In *When the Sun Darkens* this author provided a synopsis of what transpires on Earth in the Seven Seals, and now, for perspective and clarity, we briefly review the Seven Trumpet judgments of Revelation, taken from the Greek translations, in order to better understand what occurs in 2046 AD.

## Seven Trumpets

FIRST TRUMPET: ". . .hail and fire mingled with blood, and they were cast into the earth, and a *third* of the earth was burnt up; and a *third* of the trees was burnt up and all the green grass was burnt up."

SECOND TRUMPET: ". . .as it were a great mountain burning (gigantic meteorite or comet) was cast into the sea, and became a *third* of the sea, blood; and died the *third* of the creatures of the sea, things having souls, and the *third* of the ships was destroyed."

THIRD TRUMPET: ". . .and fell from heaven a star great burning like a lamp, and it fell on the *third* of the rivers, and on the fountains of the waters; and the name of the star is called Wormwood . . .and many of the men died of the waters."

FOURTH TRUMPET: ". . .and was smitten a *third* of the Sun and a *third* of the Moon, and a *third* of the stars, so that they might be darkened the *third* of them, and the day might not shine the *third* of herself and the night in like manner."

FIFTH TRUMPET: ". . .and I saw a star from heaven having fallen to the earth, and was given to him the Key out of the Pit of the Deep: and he opened the Pit of the Deep, and went up a smoke out of the pit as a smoke of a great furnace; and was darkened the Sun and the air by the smoke of the pit. And out of the smoke went up locusts into the earth."

SIXTH TRUMPET: This concerns the release of four powerful Dark Angels, Anunnaki Lords out of the Deep (Abyss) and each leads a fourth of an Anunnaki army of 200,000,000.

SEVENTH TRUMPET: Same as Sixth Seal, a time of preparation until the next seven judgments, in this case preparatory to the coming of the Seven Vials of Wrath.

In the Revelation passage here we find *eleven* things God had created in Genesis 1 reduced by a *third*, leaving 66.6% remaining of the earth and seas, flora and fauna. But the Creation account is actually a *Renovation* of a preexisting world, and the Anunnaki, called the Builders, performed the renovating. In the *Enuma Elish* text there are specifically listed *11 kinds* of Anunnaki manifestations. The first time in human history that the Anunnaki appeared among men, as the Watchers before the Flood, was in 3439 BC, when NIBIRU appeared and a *third* of the world was killed in quakes and flooding. The *last* time NIBIRU appeared, in 1314 AD, resulted with a global death toll of a *third* of humanity in the ensuing cometary fallout, meteorites, earthquakes, flooding and plagues.

The connection between the Anunnaki and comets or meteorites is quite ancient. Among the oldest written scriptures are the Vedic texts, wherein we read, ". . .hurl your crushing deadly bolt (missile) down on the wicked fiend from heaven and from the earth. Yea, for *out of the mountains* (passing comets) your celestial dart (meteorite) wherewith ye burn to death the waxing demon race . . .cast ye downward out of heaven your deadly darts of *stone* burning with fiery flame."(9) The association between the Anunnaki and mountains is clearly seen in Revelation where the Seven Kings are also called *Seven Mountains*. The mountain-evil fiend link is further evidenced in the Nordic traditions of Baldur of Scandinavia, which hold that Loki (wicked trickster god) was imprisoned below the earth in iron chains ". . .until the end can come, until the Day of Ragnarok."(10) Just as Azazel of the Hebrews was confined below the earth, so was a host of archaic traditional deities, who would one day be freed to wreak their havoc upon mankind. Loki is said to break his chains when the Ragnarok occurs, the End known to the Vikings who preserved these stories as the *Return of the Great Comets*.(11)

These Anunnaki beings are described in the Revelation as *locusts*, which are creatures known for being *destroyers of the harvest*. In this case, the harvest is *humanity*. These are the Ancient Ones of occult literature. In Islamic eschatology we find a definitive parallel that cannot be so easily dismissed.

In the Quran we read, "Lo! The day of decision is a *fixed time*, the day when the *Trumpet* is blown . . .and the *heaven is opened* and becomes as *gates*, and the hills are set in motion and become as a mirage. Lo! Hell lurks in *ambush*!"(12)

Although the eye motif was originally a symbol attributed to the Godhead as the All-Seeing Eye, it was quickly subverted and borrowed by the Anunnaki as a badge of their terrestrial authority as the *Watchers*, so prolific in the Enochian texts. The icon is found all around the world and there is evidence that the Eye above the pyramid on the US Great Seal (Dollar Bill) does not intend to portray the Eye of God but instead the emblem of the Anunnaki, vastly intelligent beings who knew that 1776 AD was not merely the beginning of the United States, but marked the 270 years until the return of NIBIRU and descent of 200,000,000 of their ranks upon the Earth in the Apocalypse. This mirrors the 270 years from 4309 BC to 4039 BC, when at the closure of Earth's ruined wandering of 270 years, the Anunnaki as the Builders *renovated* Earth 144 years before mankind was banished from Paradise in 3895 BC. Their leader appeared in the Garden as a serpent (symbolic for Interpreter) who deciphered for humanity the Tree of Knowledge of Good *and Evil*. Promising illumination, the Watchers darkened the souls and future of humanity and the Stone the Builders Rejected then had to come as Kinsman-Redeemer to raise the Temple of Man and perfect it in order to assign them to the heavenly positions the Anunnaki fell from. In 1776 AD the Bavarian Order of the Illuminati was established with much controversy and soon after went underground again. But the Anunnaki never ceased planning, for even in remote antiquity the Sumerians recognized two branches of the Watchers – the Anunnaki of heaven and those called the Igigi, stationed on earth *among men*. They are the true power and force behind the elitists like the Rothschilds, Hapsburgs, the modern Bilderbergs and other behind-the-scenes global financiers who have instigated all the major wars and socio-political disasters since 1776 AD.

In Philistia, a 6th century BC chalcedony seal was excavated from Tel Jemmeh which depicts a human-faced bird with a scorpion tail beside the Babylonian motif of the Eye.(13)  The Philistine link between the eye motif and the scorpion is not mysterious, for this is how they are associated in the older Babylonian writings. The scorpion does harm to men and wings denote divinity, the human face representing intelligence and knowledge. The scorpion is a solar symbol, and as far back as Akkad the scorpion was associated to *sunstroke*.(14)  The seal from Philistia seems to embody the idea that the Anunnaki will appear someday and *affect the Sun,* bringing harm to men. In the Babylonian *Enuma Elish* tablets we learn that the Anunnaki "made themselves masters," and were ". . .monster serpents, sharp of tooth, and relentless in attack, with poison instead of blood. . . raging monsters, clothed with terror, dragons and Lakhamu, hurricanes, raging dogs, *scorpion men*, mighty storms, fish men and rams (horned creatures), bearing relentless weapons. . . ."(15)

In 2046 AD *heaven and earth collide* and this contact opens the Gate known in the Necronomicon as IAKSAKKAK, which releases the Ancient Ones – powerful Anunnaki armies that have been imprisoned since the Rebellion in 4639 BC. Earth will be pushed toward the Sun and as the trumpet judgments unveil, virtually all flora will be burned and ruined and a third of humanity will die in the heat and disasters. The Revelation record says that unto the locusts ". . .was given power, as the *scorpions* of the earth have power."  These being the Trumpet judgments, the Anunnaki are commanded to leave all remaining plants and animals alone, tormenting *only* mankind with their fiery scorpion stings, and only those humans *not sealed* by the living God. Is this the original meaning of the *seal* from Philistia?  Did the ancient owner of the seal hope that it would provide him talismanic protection against the coming and pain of the Anunnaki?

These locust beings are described in Revelation with horse-like bodies, with the faces of men, scorpion stingers and crowns on their heads, long hair like a woman's, the teeth of a lion, breastplates (armor) hard as iron and wings. The release of these beings upon humanity is called a Woe, just as they are termed "workers of Woe," in the Babylonian texts. These physical appearances are unlike anything we have experienced. If we take these descriptions as figurative then we are provided with physically manifested demons on Earth possessing great *intelligence* (man-faced) that are of extreme *antiquity* (long hair signified age in ancient times). Incidentally, the word *demon* is derived from the root word "to know." Their horse-like bodies signify *mobility* and *warfare*, their teeth indicative of the *power of speech to harm* and insatiable voracity in harming the harvest. Their crowns and breastplates denote that these beings will rule over mankind for the specified duration of *five months*, protected from any damage man *might attempt against them*. Five months happens to also be the locust season. Their wings reveal divine origin, but their presence among men at this premature time and their limited period of five months suggests they are themselves controlled by One of greater authority than they, One that made His own earthly servants who remain faithful and *immune* to their attacks. These Anunnaki will not depart until 2052 AD, which is the subject of *Descent of the Seven Kings*. They will be mistaken for extraterrestrials that came to war against humanity and the knowledge that these are actually demons responsible for destroying the Pre-Adamic World will be suppressed.

In 1478 AD the Aztecs finished a 22-ton carved basalt disk about 10 ft. in diameter, covered in apocalyptic images and symbols called the Stone of the Fifth Sun, and the Calendar Stone. This gigantic American text is a rendition from earlier cultures like the Toltec, Maya and Olmec of the Revelation vision, the Doomsday traditions of the Hopi, the Scandinavian Ragnarok prophecies of the Vikings and the legends of the return of comets and planets from virtually universal mythos. The Stone depicts Ages that all ended in disasters that involved the *Sun* (the Ages themselves are called *Suns*). Upon the relic are pictures of Coxcox, the Mexican Noah, survivor of the Flood and evidenced in the matrix of glyphs and images are profound geometrical concepts. Bordering the artifact are numerous comets, all traveling toward a *terminal point*. The center of the disk has baffled Americanologists because it portrays the image of a deity's face with a *knife* protruding from its mouth like a tongue, the same picture we get from a casual reading of the Revelation when Christ descends from heaven at Armageddon with a *sword* protruding from His mouth (power of speech to conquer). This is the signature of the Word.

The Aztecs claim that these beliefs came from golden tablets that were brought to their ancestors by a stranger long ago.(19) These golden records are universal, many peoples in antiquity claiming to possess golden tablets with inscriptions that had been copied from off of the ancient pre-flood inscriptions once adorning the Great Pyramid. In this author's *Chronicon*, this subject is explored in more depth. What must here be acknowledged is the amazing imagery of the symbol upon the disk for the day called *4-Movement*.

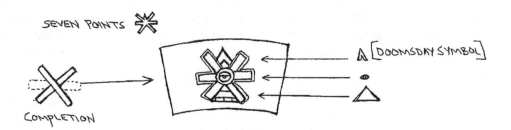

This Aztec symbol has seven terminations (7-pointed geometry produces the 52 degree angle), six of them representing *time* (bars) and the apex is a miniature Aztec *doomsday* symbol. Below the time bars is a perfectly represented *pyramid* with bricks and in the center of the image is the Eye. The 4-Movement glyph refers to something that will occur to the Earth, related by the pyramid, when the situation of *time* will change and the polestar (Eye) will be affected. Though this is an Aztec symbol, probably one they themselves little comprehended, Mayan scholars assert that the crossed bars symbolize that a *cycle has been completed*.(20) Of further significance is that the Aztecs called the city where the Stone of the Fifth Sun was found, Tenochtitlan. Not its original name, for the city is ancient, but one preserving an interesting element: T(Enoch)titlan, literally meaning Place of the City of *Enoch*. The name of this glyph as 4-Movement, a day venerated by the Aztecs in their calendrical scheme, indicates a heritage rich in prophetic association referring to a time in the future when they knew the Sun would be *moved*.

The Great Seal of the United States seems to have its predecessor in the Aztec symbol for 4-Movement, the Eye and pyramid association adopted by America in recognition that it inherited a *geographical region* that would be tragically affected. This knowledge is also apparent within the corpus of Mayan astronomical knowledge. A curious illustration of a picture in a Mexican manuscript of the Maya shows a priest watching the horizon from a temple door (left).

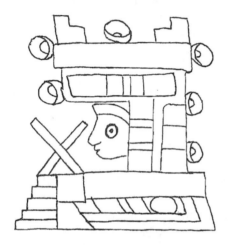

A 52-degree angled pyramid with one of its time-bars pointing into the sky like the shafts of the Great Pyramid (right).

The picture on the left was preserved in a book called *Early Man and the Cosmos*,(21) but the illustration on the right is the author's, showing how the Maya cleverly hid an allusion to the Great Pyramid of their ancestral memory beneath the picture of a temple pyramidal structure in their country. This image is virtually the same as the Aztec one. Here we find a structure atop a pyramid of five steps adorned with *seven eyes*, the priest serving as a Watcher looking for the *completion of a cycle*, which is symbolized by the crossed bars placed as a viewing instrument. In the concealed pyramid, the Eye of the Watcher lies directly at the apex of the pyramid.

The Maya, as well as the Toltec and Aztecs, all migrated from some place to the north, being North America. And it is in North America where we find more evidence of this Anunnaki-eye link – in California. The Chumash peoples of the Santa Barbara area left behind the below petroglyph, showing the descent of a *comet* and a monstrous creature, the comet having an *Eye*.(22)  The comet has ten points, a definitive pyramid number, the tetractys.

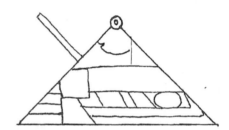

That 2046 AD is the end of a Great Cycle can no longer be questioned by the abundance of the material presented in this work. The appearance of the Anunnaki in this year is NOT what this cycle refers to, but is instead the direct *result*, the consequence, of what does happen to our planet. All the evidence mandates not only the completion of a cycle, but also the start of a whole new method of measuring *time*.

The next Archive will show just this.

*Archive 11*

## Condensed Calendar and the End of
## the Great Pyramid's Astronomical Chronology

"Behold, the days come, and the times will hasten more
than those that are past, and the years will pass more
quickly than the present years."
—*Apocalypse of Baruch* XX:1, scribe of Jeremiah

The Mayan prophecies concerning the *collapse of time* at the conclusion of their calendar is supported in the prophetic records of the Hebrews and the Revelation. The prophet Enoch before the Flood, architect of Newgrange, Stonehenge and the Great Pyramid, wrote:

"In the days of sinners the *years shall be shortened*. Their seed shall be backward in their prolific soil; and everything done on earth shall be subverted, and disappear in its season. The rain shall be restrained, and the heaven *shall stand still*. In those days the fruit of the earth shall be late, and not flourish in their season; and in their season the fruit of the trees shall be withholden. The Moon shall *change its laws*, and not be seen in its proper period (Earth and its satellite pushed out of its place). But in those days shall heaven be seen and barrenness shall take place in the borders of the great chariots of the west. Heaven shall shine more than when illuminated by the orders of night, while many *chiefs among the stars of authority* (Anunnaki) shall err, perverting their ways and works. Those shall not appear in their season, who command them, and all the classes of the stars shall be shut up against sinners (those angelic orders that usually protect humanity)."(1)

This passage in the Enochian writings, which claim they were written in the beginning for the benefit of mankind in the end, reveal clearly that the evil angelic beings will be active against humanity contemporary with a dynamic change in the movement of the Earth. The Anunnaki Homeworld ascends from the Deep, below the solar system, and crosses over the ecliptic, fulfilling its role as Planet of the Crossing as it traverses the ecliptic of the Sun-Earth orbits, nearly colliding into our own planet. This single event in 2046 AD is the subject of the Hopi prophecies of the coming of the Dark Star and other American traditions concerning the lost or Black Star, the Black Sun, initiating the *Fifth Sun* (fifth time the motion of the Sun *changed*). It marks the return of the "fearful constellation," foretold by the priest-prophet Ezra.(2)

91

Ezra wrote that the Sun will appear and shine at nighttime and the Moon will appear *three times in a day*.(3)  Wild animals will migrate in massed confusion and menstruous women will give birth to monsters,(4) genetic aberrations. Women will give birth to children in *three or four months*, the infants shall live and be raised up, talking at the age of *one*.(5)  Natural springs and rivers will initially not flow for three hours.(6)  The first, second and third trumpets involve meteoric rain, asteroid and comet impacts that burn up a third of Earth and destroy fresh water lakes, rivers and reservoirs, the fourth trumpet being the *result* of the close proximity of NIBIRU:

> ". . .and a third part of the Sun was smitten, and a third part of the Moon,
> and a third part of the stars; so as the third part of them was darkened,
> and the day shone not for a third part of it; and the night likewise."
> —Revelation 8:12

The immense gravitational attraction of the mass of NIBIRU will pull on the Earth and draw our planet into a closer orbit toward the Sun, causing the intensified heat that burns up a significant portion of the planet in Apocalypse. Further, Earth will be pulled *off the ecliptic* and back onto the ecliptic of the Dark Star (former Daystar of lost binary NIBIRU still travels). The Sun itself may undergo a pole shift, aligning itself with the Galactic Plane, tossing all the planets from Mercury to Neptune into chaos as they restabilize into newer orbits. This will reveal the Original Zodiac, which was not in the present ecliptic but found its origin in the stars and constellations spanning the Dark Star's ecliptic plane. The prophet Isaiah confirms this as a future condition of earth when he wrote, ". . .the earth is utterly broken down. . . moved exceedingly. The earth shall move to and fro like a drunkard, and shall be removed like a cottage. . . and it shall come to pass in that day that the Lord shall punish the *host of the High Ones* that are on high (Anunnaki) and the kings of the earth upon the earth. . . the moon shall be confounded, and the Sun ashamed (a reference to covering its face) when the Lord of Hosts shall reign in Mount Zion. . . ."(7)

The imagery of the Revelation record employs symbols and concepts used by the ancients to describe NIBIRU. In 1915 BC when the Anunnaki planet passed Earth and was viewed by the Babylonians, they preserved in their cuneiform tablet texts that it appeared as a giant, dark, multi-headed dragon, and they called it Tiamat, the personification of Chaos. The Egyptians referred to NIBIRU as the celestial monster, Typhon. In Hesiod's *Theogony* it is related as huge *dragon heads* having black tails, and through the course of history from China to ancient America, comets have since then been associated to dragons. This concept remains unchanged in the Revelation text, which also describes NIBIRU approaching the Earth.

> "And a great sign was seen in heaven; a Woman invested with the Sun
> and Moon under her feet. And on her head a crown of twelve stars, and
> being pregnant, she cried forth travailing and being pained to bring
> forth . . . And another sign was seen in heaven, and behold! a great *fiery-
> red dragon*, having *seven heads and ten horns*, and on his heads, seven
> diadems. . ."
>
> —Revelation 12:1-3

This Woman is the constellation Virgo, and the *only* way to cause the Sun to be *under* the feet of this zodiacal constellation is for our own planet to be *removed* from the ecliptic, to be placed far *above* the ecliptic plane so as to make the Sun below Virgo. What this implies, sequentially, is that NIBIRU will ascend into the inner system from directly underneath Earth, push our world onto the Dark Star's ecliptic plane with it, and with Earth shoved upward and out of the way, we will be able to first view Virgo before seeing NIBIRU as it passes our planet on its 60-year journey over the Sun.

As the Maya knew, *time will collapse* into a newer, more accelerated system. Earth's new position closer to the Sun will increase its rotational spin-rate by a *third*, from a 24 hour to a 16-hour day, and the year will be abbreviated from the present to a third of the original year of 360 days, to *240 days*. The orbit is quickened, obviously because the planet's distance from the Sun is significantly shorter, and the rapid rotation increases our ability to see the Moon more often than is presently possible. This is how both day and night together can be reduced to a *third* without lengthening one or the other.

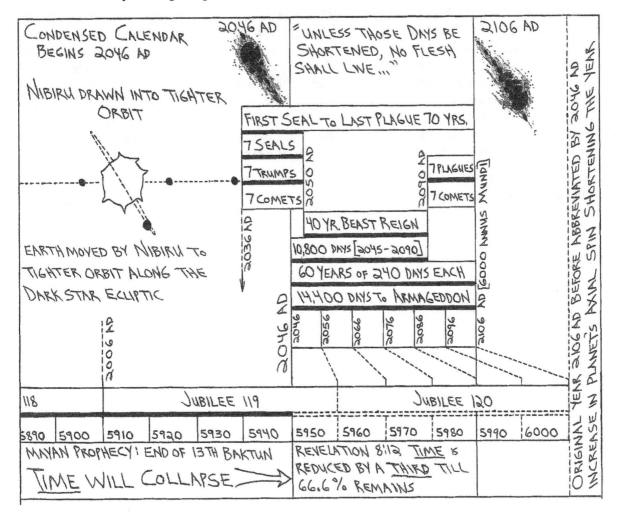

This is a deliberate act of God, for Christ said, ". . .unless those days be *shortened*, no flesh would be saved." This is the Last Days Quickening. The new orbit is the precise opposite, the reverse of what transpired in 4309 BC when the Pre-Adamic World was destroyed and Earth was pushed out of its place in a solar-system wide disaster. The initial explosions of the Daystar that imploded, becoming the Dark Star, sent Earth plummeting through space for 270 years until reaching the present Sun and intrusively initiating its own orbit between the orbital belts of Venus and Mars where it remains a mathematical anomaly. This 270 years, according to the Sumero-Babylonian records, read that ". . .*long* were the days, years were *added*." In 4039 BC Earth began rotating and thawing out as the Builders (Anunnaki) set to the task of renovating Earth. This was a *collapse of time*, and the new rotation and orbit initiated a new Astronomical Chronology that was encoded in the dimensions of the four cornerstones of the Great Pyramid and discovered in the early 20th century AD by David Davidson as ending in 2046 AD (he was off by a year). The Maya preserved a system that counted down to this end of Earth's Astronomical Chronology, but their countdown began in 3113 BC. Now, in 2046 AD, the planetary situation is *reversed* to ". . .*short* were the days, years were *taken*. . ."

Thus, everything God had made in the beginning, from the vantage point of Earth, was reduced to a third in the *end*, even *time*. This alteration reveals why the Antichrist changes *times and seasons*, as prophecies in the Scriptures dictate. It will also accelerate and increase the violence of the Apocalypse, for the Revelation unveils that the Dragon (Lord of the Anunnaki) will be frustrated because ". . .his time is short." This also explains why the Scriptures read that Christ ". . .comes as a thief," in the Last Days. He literally steals *time* from the Anunnaki.

The chronological effect of the Condensed Calendar is an abbreviation of the final 60 years from 2046 to 2106 AD (6000 AM), to a *40-year* period, the elimination of a third of this time resulting in the actual loss of 20 years. The 40 years of 240 days each is exactly 14,400 days (144 x 100) from 2046 AD to the year 6000 Annus Mundi. How this acceleration will affect human longevity is not known; however, what is abundantly clear is that at the time of this writing, there are hundreds of millions of people living who will still be around to witness these events. In fact, the entire generation born from 1986 AD and afterward will potentially still be alive for the return of the Chief Cornerstone in 2106 AD, for the Condensed Calendar abbreviates the 120 years to 100 years. Nonetheless, the 2046 AD event and start of this new Condensed Calendar is only a few decades away from the composition of this work.

As the subject matter of this Archive concerned the condensing of time itself in the near future, so too will the next Archive remain as brief as the topic it expounds.

# Archive 12

## Stonehenge and the Chronometry of Original Time

"So men began to live and understand the destiny assigned to them
by the course of the circling gods . . . leaving great memorials of their
work on earth."   —Corpus Hermeticum (1)

In this author's work entitled *Lost Scriptures of Giza* is profiled the history and evidence that the prediluvian patriarchal prophet-king Enoch designed the Great Pyramid, which was completed in 2815 BC, the year 1080 Annus Mundi. This sage was born in 3273 BC (622 AM) and became the most famous of the Sethite kings. He was the son of Jared, who was specifically named in commemoration of the *descent* of the Watchers before the Flood. His father witnessed against the Anunnaki and their Nephilim offspring, a ministry Enoch inherited, for in 3233 BC when Enoch was 40 years old he too began his ministry of witnessing against the Giants and their Anunnaki fathers about their depravity and the coming of divine punishment for both men and angels. This history is provided in *Chronicon*.

But our learned king and prophet inherited something far more important to us today than an antediluvian ministry. As the Enochian records convey, he also learned the *future*. When his father was 40 years old in the year 3395 BC, which was the 500th year of the Annus Mundi (and pre-corrupted Hebraic), Jared received a vision from the Lord concerning the *end* of the reign of the Watchers and the return of Earth to paradise, the conclusion of sin and rebellion and the dark oppression of angels over the sons of men. God informed Jared that this timeline would be concluded in exactly *5500 years*.(2) This means that Jared, and then his son Enoch, *knew* the timeline of God, which translates into our own calendar today as *2106 AD* (6000 AM).

When Enoch was in his 70th year of ministry, his prophetic messages took on another form. With his books written at instruction of the Spirit, he now employed the science of the Builders and began his *chronometry* projects, literal stone megalithic prophecies immortalized within the dimensions of permanent stone. This year was 3163 BC (732 AM), and the first project initiated by him was what is now called *Newgrange*. This dating is confirmed by astrophysicist Tom Ray of Dublin Institute for Advanced Studies, who discovered that sunlight would have penetrated the back vault wall, deep within the artificial mount of 5150 years earlier, at the exact moment the Sun emerged at the vernal equinox. Amazingly, this date is 732 Annus Mundi, paralleling the 732 years of NIBIRU's sub ecliptic orbit, 366 years to aphelion and 366 years back (732).

Newgrange in Ireland is a gigantic earthworks site with a megalithic tunnel leading into a cruciform vault, which employs the same angular dimensions as Stonehenge. Its surrounding standing stones form the same 135-degree angle as Stonehenge's station stones. As these monuments display the same features, one is buried under tons of earth while the other is open to the sky, and 8 total monoliths in the 135-degree geometry provides us with the calendrical and geometrical identity of the Great

Pyramid, for 135 x 8 is *1080*. More on this later. The similarities between Newgrange and Stonehenge are detailed in *The Stone Angle*, by Jack Wun.

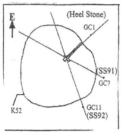

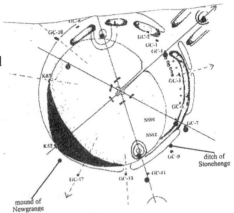

The Newgrange complex was finished by the vernal equinox and the builders, along with their architectural prophet, then went on to southern England and erected the famous and enigmatic Stonehenge, the focus of this incredible Archive. Many theories have been postulated about the purpose of the site, cultists seeing mystical application while astronomers read into the stones various arbitrary alignments involving the stars or constellations. Culturalists attribute the site to Druids having erected a wonderful temple, while more exotic theories serve to show that the site was built for an extra-terrestrial purpose. None of these ideas approach the truth. With a hundred-thousand stars in the heavens it would be surprising not to find some alignment that a measure of significance could be attributed to, but the astronomers of today even using advanced computers have yet to publish anything convincing, substituting their flashy graphics and photographs for their lack of knowledge, and further increasing the mystery of the site. It was Sir Norman Lockyer's research in 1901 of Stonehenge that began the general belief that the ancient megaliths were associated to the stars, a concept he espoused in his *Stonehenge and Other British Stone Monuments*.

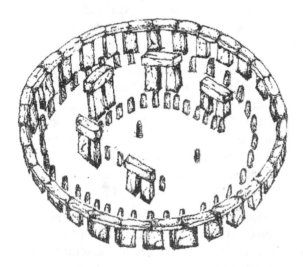

The language of ancient prophecy was *geometry*, and the key to interpreting geometrical mysteries encoded within the marvelous structures left behind is by simple arithmetic of its *chronometry*. A structure's chronometry is determined by its' *dimensions* and an object's dimensions are simply its planes (faces), corners and angles. These facts demonstrate that the architecture of the ancients built prior to the Great Deluge encode many secrets and facts about *time*. The formulae for determining a structure's chronometry is so simple that now, over 5000 years after these monuments were built protecting these secrets, we can interpret them with ease. Further, as we will see in the following Archive, the same formulae used to decode Stonehenge and other monuments is the same used today to decode *crop circles*, the *signs in the earth* prophesied to come in the Last Days. The science of chronometry may have also given rise to the mystic knowledge of the Enochian alphabet, first brought to the public's attention by the famous western occultist John Dee, who claimed that this was the language of *angels*, a language constructed from strange *angles*. Tracy Twyman notes that there may be a further connection between the English word *angel* (which means *messenger*) and the geometrical word *angle*. Interestingly, English is derived from the German family called the *Angles* (Anglo-Saxons).(3) This author merely makes note of these parallels because Stonehenge is today standing in the land of the Angles.

The site was erected in three distinct phases, Stonehenge I, II and III. This fact alone has engendered much scholarly confusion.

Stonehenge I (Outer Sarcen Ring)

30 sarcen pillars

30 sarcen lintils

1.  Megalithic sarcen ring forms 360-degree circle of contiguous stone, identifying the archaic 360-day year.

2.  Ancients divided the 360 degrees of heaven into 12 regions (Zodiac). The 30 lintils (*raised in air* signifying heaven) x 12 is 360, representing the *astronomical year.*

3.  The 30 massive pillars are the *foundation* supporting the raised lintils, each pillar having 4 sides: 30 x 4 is *120. . .* the *FOUNDATION* number (a third of 360).

4.  Each 6-sided stone has 8 corner angles. The 30-sarcen pillars have 240 total corner angles, but only *120 are visible* (120 buried underground).

5.  Each of the 6-sided lintils was attached end-to-end with advanced tongue-in-groove locks, eliminating 2 planes on each stone for a total of *120 visible planes.* Each lintil maintains a spherical curvature forming a perfect raised ring, unlike anything else found among ancient megalithic architecture.

6.  In antiquity holy men were regarded as *pillars* (pillar of the community) who supported heaven (*men* means stone: i.e. menhir) as in the Book of Revelation where the redeemed among men shall become ". . .*pillars* in the temple of God," a pillar being the symbol of a *repository of divine knowledge.* The number of MAN is 6 in the LAST book of the Bible (Revelation) and in the FIRST book (Genesis), mankind was created on 6th day. Thus, the 60-sarcen pillars and lintils, each having 6 planes (60 x 6 = 360), links the concepts of MANKIND and TIME.

7.  Total sum of corner angles and planes of the 30 lintils in the outer ring was 420, however, only *360 were visible* due to their attachment end-to-end.

8.  The 120 visible corner angles of the sarcen pillars combined with the 240 visible joints of the lintil corner angles are *360 total visible* angles in the outer sarcen ring of Stonehenge.

9.  The outer sarcen ring's PLANES and CORNER ANGLES combined together in sections 6, 7 and 8, above, demonstrate the sum of 360 three different ways, for a total of *1080* (360 x 3): GREAT PYRAMID FINISHED 2815 BC, OR *1080* ANNUS MUNDI.

10. Total dimensional computation of sarcen ring's pillars and lintils (480 angles and 360 planes multiplied against each other) is *86,400,* the precise sum of *seconds in a day,* the second being the lowest ancient denominator of *time* and 864 is the *foundation of time* number in seconds, minutes, hours, days, months and years. There are many crop circles formed near Stonehenge that employ geometrical dimensions that mathematically equate to 864 and its variants (see Crop Circle Calendrical Codes).

11. Sarcen ring's 30-pillared divisions x 360 produces 10,800.

12.   Sarcen ring's total of 60 stones (pillars and lintils) x 360 produces 216,000, the exact sum of days in a 600-year period (Great Year or Anunnaki *NER* period); the reoccurrence of 60 and its multiples denotes a connection to the Sumerian sexagesimal system.

CURSED EARTH CALENDAR [STONEHENGE CHRONOLITHIC SYSTEM]

PLATE II.

PERSPECTIVE OF VIEW STONEHENGE AT TIME OF SUNRISE OF SUMMER SOLSTICE.    1680 B.C.

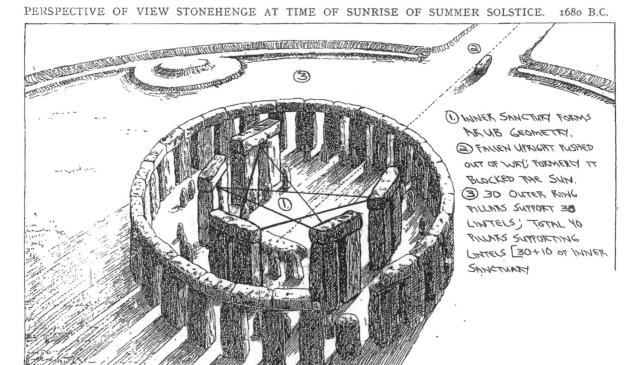

(1.) INNER SANCTURY FORMS ARUB GEOMETRY.
(2.) FALLEN UPRIGHT PUSHED OUT OF WAY FORMERLY IT BLOCKED THE SUN.
(3.) 30 OUTER RING PILLARS SUPPORT 30 LINTELS; TOTAL 40 PILLARS SUPPORTING LINTELS [30 + 10 OF INNER SANCTUARY

*Drawn by A. C. de Jong*

## Conclusion

The outer sarcen ring of Stonehenge embodies the concept of Original Time in the beginning of Men, the 360-day terrestrial year synchronized perfectly with the 360 degrees of the celestial sphere. Not coincidentally, the outer circular embankment around the site (ditch and wall) has a diameter of 360 feet. (4)

The sarcen ring identifies the calendrical units by which the rest of Stonehenge (the Trilithons) and its later additions (Bluestones) employ in its calendrical messages. These units of *time* encoded within the dimensions of the outer sarcen ring are:

| 1 day | (86,400 seconds) |
| 60 days | (86,400 minutes) |
| 120 days | (and years) |
| 360 days | (8640 hours; and years) |
| 30 years | (10,800 days) |
| 600 years | (216,000 days; Great Year) |
| 1080 years | (Annus Mundi year Great Pyramid finished in 2815 BC) |

The connection to the Great Pyramid is made more profound by the fact that Stonehenge lies near 51.52 degrees latitude,(5) and the sloping angle of the Great Pyramid's four faces is *51.52 degrees*. The sarcens are a hard-grained sandstone rock emblematic of *men* made from *earth* (sand, dirt), as opposed to the chemically altered metamorphic or igneous rock. The outer sarcen ring is 13 feet high, and the 10-foot long lintils placed above them are *six tons each*. The sarcen pillars that support them are six feet wide and 3 feet thick and weigh *25 tons each*. Just this outer ring, all connected together in amazing tongue-in-groove links, with a circular curvature in the stone *found nowhere else on earth*, weighed a total of *930 tons*. . . specifically designed to endure for *millennia*.

## Stonehenge I (Trilithons)

These inner monoliths are virtual rock giants, and the best-preserved features of the Stonehenge complex. They are sarcen stones, 24 feet high and up to *45 tons*. The Trilithons are formed of 10 pillars, forming a horseshoe topped with five lintils (see facing page sketch). These lintils are not connected to one another as they are in the outer sarcen ring, giving one the distinct impression of five great Ages.

The Five Trilithons are like five portals, forming the Inner Sanctuary of Stonehenge. Their distribution forms an ancient geometrical symbol for the Creator, derived in antiquity from the belief that the Eye of God was the polestar, the only star in the heavens that did not move, with all stars encircling it since the Earth's axial line pointed directly at it. To the ancients this gave off the impression that all the stars in the sky worshipped the Polestar. The Trilithons first form this five-point geometry, but the star symbol itself is formed also by the four Station Stones (see right) outside the sarcen ring, and the long line, below, intersects within the Trilithons to form this ancient sign for God.

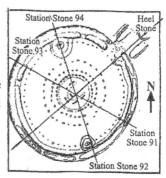

The total corner angles of all 15 trilithons are *120* (15 x 8). Because the trilithons depict a pentagram with 5 corner angles, we find produced the sum of *600* (120 x 5). 600 years was a Great Year, also known as an Anunnaki NER. The Great Flood of 2239 BC was the 600th regnal year of the Anunnaki Kings who began to rule in 2839 BC, 600 years after they descended to Earth in 3439 BC.

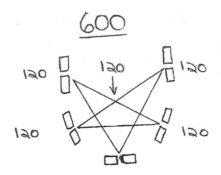

The pentagram forms the Sumerian AR UB sign, the Plough of God, a *comet* sent to *break up the ground* because His harvest (humanity) has ceased bearing fruit . . . This ☆ is same sign as the Kali Yuga, or Black age.

 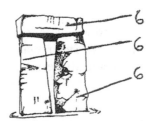

The sum of 600 is verified by fact that the total sum of corner angles in the sarcen outer ring and sarcen trilithons is *600* (75 sarcens x 8). By extension, the 5-point geometry infers the greater period of 3000 years (600 x 5).

The pentagram of 5 equidistant points is formed by ten angles of *108 degrees* (also reflected in 10 trilithon pillars). This 108 x 10 sum is *1080*, the exact year the Great Pyramid was finished in 2815 BC (1080 AM). When 3-dimensionally depicted, the pentagram forms a *pyramid* (right).

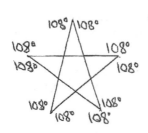 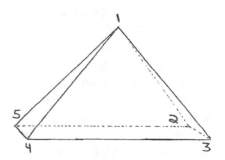

Another formula producing the sum of 108 and 1080:

6 is the number of MAN (stone)

6 planes on each trilithon stone

3 stones is a trilithon

---

6 x 6 = 36

36 x 6 = 216 (108 + 108)

As each trilithon becomes 216, five trilithons add up to *1080* (216 x 5). Conceptually, each of the 10 pillars of the 15 trilithons represents 108.

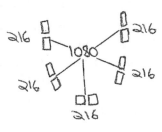

Further, the 10,800 produced by the outer sarcen ring are replicated here in the sarcen trilithons. This is found in multiplying the 120 corner angles by the 90 planes (15 x 6) to get *10,800*. This is evidence that the *sarcens* (outer contiguous ring and inner trilithons) were the ORIGINAL STONEHENGE, their dimensions complimenting each other.

The repetition of the number 6 (6 x 6 x 6) provides us with a clue as to the importance of this complex to both God and man. The cube of 6 forms the gematria for DBIR, or MOST HOLY PLACE, which represents that the *stone* (symbol for man) is the container of something of epic magnitude, the Holy Spirit.(6)  This is the secret message underlying the difference between Stonehenge I and Stonehenge II, the introduction of the Bluestone ring and horseshoe. As we will see in this Archive, the Bluestones represent the *Anunnaki*, and these primordial beings were *offended* that their own creation (mankind), when they acted as the Builders who performed the work for the Creator, would contain within them a portion of the Spirit of the Living God, despite the fact these beings (mortals) were made of mud.

Because the Trilithons are not connected as the outer sarcen ring by lintils, each trilithon has 18 faces (combined planes of all three stones) and 24 angles producing the sum of 432 (144 x 3 / 108 x 4). As there are five Trilithons, this is what is unveiled:

1.  432  gematria for *world, habitation; all nations*

2.  864  Foundation of Time (seconds, minutes, hours, days, months, years)

3.  1296  144 x 9 / 108 x 12

4.  1728  144 x 12 / 108 x 16

5.  2160  1080 x 2; one House of the Zodiac (216 x 10)

Once we add the 360, conceptually realized in the outer sarcen ring to the Trilithon message, we get 2520 (360 x 7), resulted by adding 360 to 2160. The prophetic writing on the wall in the Babylonian palace the prophet Daniel interpreted was MENE MENE TEKEL UPHARSIN, meaning literally NUMBERED, NUMBERED, WEIGHED, DIVIDED, while also being gematrical messages equaling 1000, 1000, 500 and 20, or *2520*. The fall of Babylon in that very night to Persia in 537 BC began a 2520-year countdown to the 1984 AD Iran (Persia)-Iraq (Babylon) War.

In this author's prior work, *When the Sun Darkens*, is found a detailed analysis of world history as it unfolds in 414-year periods (207 + 207), a system called the Cursed Earth Chronology which began with the ruin of the Pre-Adamic World in 4309 BC and ended 6210 years later in 1902 AD (5796 AM), beginning a 144-year countdown to 2046 AD. This system is specifically referenced in the Stonehenge I chronolithic dimensions of the complex. The Cursed Earth system involves the planet Phoenix *darkening the Sun*, as well as transiting comets, as well as *impacts*. The 6210-year longevity of the Cursed earth system is referenced by taking 207 x 30 sarcen pillars (6210). The 1902 AD end-date of this Cursed earth system serves to show how the Stonehenge Calendar merges into the beginning of the Giza Course Countdown that began in 1902 AD and counts up through the monument 203 levels of masonry until the year 2106 AD, when the Chief Cornerstone descends in 6000 Anuus Mundi to vanquish the Anunnaki.

### Stonehenge Cursed Earth Calendar

Time is governed by the motion of the *planet* around the Sun (typological symbol for God). The five Trilithons embody the concept of an astronomical planetary clock that will indicate the coming of *judgment upon mankind*. The Trilithons are five *planetary gates* for Mercury, Venus, Mars, Jupiter and Saturn. The Great Trilithon's height down to the floor entrance of the sarcen outer ring is 23 degrees, the angular tilt of our own planet, which is the result of judgment. The number for the concept of *judgment* is 9. The product of 23 x 9 is *207*. The sum of 207 is the base calendrical period in the Cursed Earth Calendar.

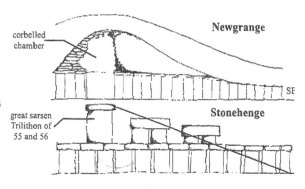

The designer of Stonehenge (Enoch) was aware of the coming Deluge that would destroy the world and result with the 23-degree tilt of the planet (broken planet). The monument was designed to convey the message that the *planets* would warn mankind when the final epoch (207 years) began. This occurred in the year 2446 BC (1449 AM), as recorded by Chinese astronomers when, *207 years* before the Flood (2239 BC), the entire world witnessed *all the visible planets* in a straight line, like a ladder into heaven.(7)

The 10 trilithon pillars add up to 2070 years (207 x 10). This is important, calendrically, because the Great Flood of 2239 BC was exactly *2070 years* after the total destruction of the Pre-Adamic World in 4309 BC. 207 years is the base sum calendrical unit of the Cursed Earth system, the prime epoch measuring *414 years*. The Trilithon geometry of the pentagram x 414 is 2070.

The Trilithons each produce the sum of 432 years, but applying the *same* mathematical formula by which we interpret *crop circles* that appeared next to Stonehenge in 1996 (Julia Set), we arrive at the sum of *414 years*:

24 angles x 18 faces of trilithons is 432
432 total minus the faces is *414*

The Julia Set next to Stonehenge on Salisbury Plain is a 1996 crop circle formation that encodes the sum of *4140. . .* ten Cursed Earth periods.

The entirety of the Cursed Earth Chronology is 6210 years (414 x 15 trilithons) or 2070 x 3 years from 4309 BC, when the Pre-Adamic World was completely destroyed, to 1902 AD (5796 AM), being Year One of the Giza Course Countdown of levels of bricks to year *6000 Annus Mundi* (2106 AD). The 144th year of the Giza Count is 2046 AD, when a comet/asteroid will destroy a *third* of Earth.

The mathematical, geometrical and chronometrical relationships between Stonehenge and the Great Pyramid are many, which is to be expected because the architectural prophet Enoch designed them both. The 2070 years of Stonehenge becomes the year *2070 Anno Pyramid* (2070 years from completion of the Great Pyramid in 2815 BC). This is 745 BC (3150 AM) when the Tribes of Israel were first deported out of their lands of inheritance by the Assyrians and scattered, later migrating into Asia and Europe. This began an EXILE period of judgment for *2520 years* (360 x 7) until the reemergence of the descendants of the Lost Tribes would convene and begin a Last Days empire, prophetically known as Ephraim and Manasseh, the 13th Tribe, the *Empire of Adoption*. These 2520 years ended at *1776 AD AMERICAN REVOLUTION* (5670 AM). This final empire of Israeli descendants has *forgotten who they are* (Manasseh means *forgetful*) and they will dominate for 270 years until a *comet* destroys their empire in 2046 AD. This judgment completes the design of the Creator in putting all of Israel back into the lands of their forefathers, for prior to comet impact in North America, there will be a mass *exodus* of American refugees who will by then understand that they are genetically linked to ancient Israel, and others will understand that even without blood kinship, they are *adopted through faith* into the people of God.

414 years before 1776 AD was 1362 AD, when the Norse and Goths explored North America and left the Kensington Stone.

The Stonehenge "sign" of the planetary gates provides us warning in our own generation. In 1940 AD a stunning five-planet alignment of Mercury, Venus, Mars, Jupiter and Saturn formed a ladder in the sky as recorded by the Chinese and found in ancient Egyptian symbolism. This indicates that JUDGMENT IS COMING. That this judgment is coming to "break up" (Sumerian AR UB) America is found in the fact that *207 years* prior to the 1940 AD planetary alignment was the year 1733 AD, when the 13th ORIGINAL COLONY of America was founded. Georgia.

### Trilithon Calendrical Unit

414 (4140)
600 (3000/6000)
1080
2070
2520
3000

### Perimeter Mound

This is a 360-degree circular embankment built with the sarcen ring and trilithons that make up Stonehenge I. It has two breaks, when lines are made to intersect at the center of the complex, they form a *135-degree angle*. This is another similarity with the Great Pyramid (135 x 8 = *1080*) seen priorly in the Station Stones and in Newgrange's Standing Stones.

Double break in the embankment implies "year" will be *broken twice* (713 BC and 2046 AD).

The sum of 120 has reappeared over and again in Stonehenge, a sum biblically significant. Noah was warned

*Courtesy of American Museum of Natural History*

83. Ladder of planets as they appeared soon after sundown, February 28th, 1940. Reading downward they are Mars, Saturn, Venus, Jupiter and Mercury.

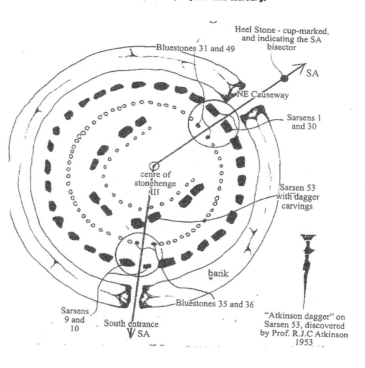

120 years before the Flood that the Earth would be destroyed. Moses lived 120 years, fulfilling a prophetic destiny. The number 120 is 40 x 3, the sum of 40 years being a *generation of man*. It is also emblematic of spiritual maturity, and we note that both Jared and Enoch were 40 when they began their ministries, and Moses was 40 when he had his first encounter with God in Midian. The intent of the Station Stones was to form a rectangle, and geometrically, the number 40 translates into a 5 x 8 rectangle, which itself conceals a latent geometrical clue underlying the science of Stonehenge.

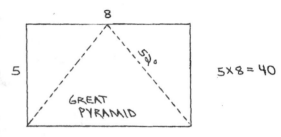

Stonehenge was specifically designed to target 5796 Annus Mundi (414 x 14), the year the Giza Countdown began 144 years prior to 2046 AD. The 15 megaliths that form the three Trilithons each represent a 414-year period in the 6210 year timeline to 1902 AD, the first 414 years being prior to the start of the Annus Mundi calendar, when in 4309 BC the Earth was destroyed in a Pre-Adamic cataclysm caused by the Anunnaki.

The overt geometrical message of the trilithons is that of *comet impact*, for the pentagram formed of intersecting lines connecting the five portals forms the Sumerian AR UB symbol explained in *Descent of the Seven Kings*. Further associating Stonehenge I to the Cursed Earth Chronology, demonstrated in *When the Sun Darkens*, is the existence of the terribly misunderstood monolith called the Slaughter Stone, a sarcen giant being 20 feet long, 7 feet wide and 5 feet thick, which originally *stood upright*.(8)

Presently, this enigmatic colossus lies beside the Causeway outside Stonehenge proper. Only because it has been *moved* can Stonehenge serve as an excellent place to observe the summer solstice. Because archeologists cannot fathom why the entire magnificent Stonehenge complex would be erected only to have a single monolith block the path of the sun's light into the Inner Sanctuary, thereby annulling what they believe was the intent (to view the sun), the Slaughter Stone is totally ignored. It doesn't fit their interpretation; therefore it remains omitted from serious scientific consideration. Perplexed, but not silent, they concocted the story, without evidence, that this monolith was for sacrifices, naming it the Slaughter Stone. Case closed.

In 2239 BC the *Sun darkened* and the planet was virtually destroyed in the 3000th year of the Anunnaki Chronology (begun 5239 BC), and 2070 years after the Pre-Adamic world met an even worse fate. In 1687 BC the *Sun darkened* and a global quake destroyed many megalithic cities around the world. The *Sun darkened* again 3000 years later in 1314 AD when NIBIRU occulted it in transit, attended by plagues, quakes and disasters. In the year 1135 BC the *Sun darkened* and again in 583 BC when Thales predicted it, as well as in 1764 AD when astronomer Hoffman observed the phenomenon – EVERY one of these events occurring on a specifically predictable timeline – the Cursed Chronology. This is the secret of the Slaughter Stone being all to itself out on the Causeway. When placed back in its original position, this gigantic block of stone will *darken the Sun*. In fact, this stone is also called the Heel Stone, anciently referred to as Freya seal, or *Day of the Sun*.(9)  Long ago, Druids or some other visitors to the site toppled the monolith to the edge of the causeway, not comprehending its significance. Or perhaps it was an earthquake.

Inside the Trilithon's sanctuary is located the Altar Stone, a sandstone of green micaceus rock. It is 16 feet long, 3 and-a-half feet wide and one foot, nine inches high, weighing 6 tons. The combined weight of the Stonehenge I sarcens (outer ring, lintils, trilithons, Slaughter Stone and Altar Stone is approximately 1400 tons, possibly more. An outer embankment encircles it, 6 feet deep, with the rise being 6 feet high. Stonehenge I, erected in 3163 BC (732 AM), consists of *all* sarcens:

30   sarcen pillars (outer ring)

30   sarcen lintils (outer ring)

10   sarcen pillars (trilithons)

 5   sarcen lintils (trilithons)

 4   Station Stones (sarcens inside Embankment)

 1   Altar Stone (sarcen)

 1   Slaughter Stone (sarcen)

Combined total is *81 sarcens*, with only 80 of them inside the Stonehenge Embankment because the Slaughter Stone is located outside, along the Causeway. Despite the apocalyptic motifs seen in the timelines associated with Stonehenge geometry, it can almost be maintained that Stonehenge also serves as a type of *Messianic* timeline because of its terminal link to 1902 AD, a calendar by itself that through *masonry* marks the years to the descent of the Chief Cornerstone.

The date for Stonehenge's beginning in 3163 BC is evidenced in another compelling way. Alexander's scholar-friend Callisthenes found a Babylonian record claiming that their history spanned back 34,000 "years."(10)  But this is a lunar code to be divided by 12 "months," which gives us 2833.3 years. And counting backward, this 2833.3 years from 330 BC gives us the precise year of 3163 BC, Enoch's 70th year of ministry when Stonehenge I was completed.

With Stonehenge I complete we now enter into Stonehenge II, and the incredibly fascinating crypto-prophecies of modern Crop Circles.

## Archive 13

## Stonehenge II, Crop Circles and the
## Anunnaki Necklace

"...And I will show wonders in the heaven above, and *signs in
the earth beneath...*"

—Acts 2:19

The 6000-year timeline of the Curse upon humanity is measured out in exactly 120 Earth generations known as *jubilees*, a 50-year period. Stonehenge I was completed in 3163 BC and remained unaltered for 50 years until Enoch was instructed by God to add *another* ring and horseshoe within the complex. These new additions are pillars only, having no lintils, and they seem to mimic the originality of Stonehenge I. But these are not sarcens, and they appear to be architecturally less significant than the initial gigantic sarcen ring and trilithons of Stonehenge I.

These newer pillars are *bluestones*, six feet high, 3 feet wide and one and a half feet thick, made of *igneous rock*, which is produced under intense heat. Symbolically the sarcens represented the nature of humanity, made from soil, mud and sand, but the bluestones, by their color, signify created beings not from Earth, but from the *sky*, composed not of terrestrial materials, but from *divine fire*. The Presli Bluestone site is aligned with Caldy Island, Glastonbury, Lundy Island, Boscastle and Castle Dor, all forming a triangle with *Stonehenge* as the apex and Lundy Island as its central base, exactly *108 miles* away.(1) This appears to copy the geographical fact that from the Great Pyramid to the islands of the coast of Egypt, due north across the Delta, is 108 miles.

The addition of the bluestones concerns a prophetic message introduced into man's calendars of the Last Days return of the Anunnaki Seven Kings, the same regents that caused the social conditions that eventually led to the Great Flood. This subject does not lead us further astray from our principle topic of NIBIRU. Both Stonehenge II and the recent phenomenon of Crop Circles are interrelated, and both serve to demonstrate knowledge of the Anunnaki and NIBIRU.

### Stonehenge II: Addition of 79 Bluestone pillars

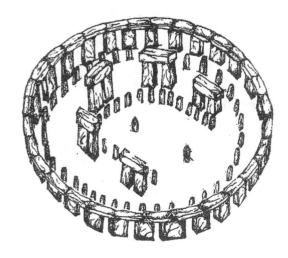

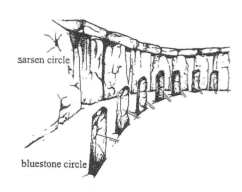

*Bluestone Ring*: 60 bluestone pillars without lintils were later added to Stonehenge, forming a 360-degree ring inside the megalithic sarcen ring of Stonehenge I.

*Bluestone Horseshoe*: 19 bluestones were added inside the Trilithon horseshoe.

Note: While the large sarcens are symbolic of man and his relationship to *time* and the earth, the bluestones imply the introduction of something *celestial*, the blue-tinted heavens, or if we consider Euphratean beliefs, blue stones were also representative of the *Anunnaki*, whose symbol was that of a *necklace* belonging to the goddess ARURU.

By adding the 79 bluestones to Stonehenge, the total number of standing *pillars* is now *120* (foundation: ground). This is 30 sarcen ring pillars, 10 sarcen trilithon pillars, a 19 bluestone horseshoe, 60 bluestones in ring and central Altar Stone.

The two rings of Stonehenge II (sarcen and bluestone) add up to a total of *120 stones* (30 sarcen pillars, 30 sarcen lintil, 60 bluestones).

Stonehenge I of the sarcens represented MANKIND ON EARTH enjoying harmony and balance. But Stonehenge II introduced celestial elements that DO NOT BELONG among MAN. Enoch was told the future of the world and the Anunnaki that were already afflicting his kingdom. Newgrange and Stonehenge I were erected in 3163 BC, Enoch's 70th year of ministry against the Anunnaki. And likewise, Stonehenge II was also designed by this architectural prophet.

Stonehenge I was a total of 81 stones and Stonehenge II added 79 bluestones for a total of *160 stones*. The number 160 was the signature of Enoch, for Stonehenge II was finished in the *160th* year of the *life of Enoch* the Prophet (born 3273 BC).

Confirmation is found in the fact that the 160th year of the life of Enoch was exactly *3113 BC*, the start date of the sorely misunderstood *MAYAN LONG-COUNT Calendar*! Stonehenge II with its introduced bluestones, verify, as we will now see, the connection between Enoch, Stonehenge, the Anunnaki and the Mayan Long-Count.

Stonehenge I's 40 sarcen pillars (30 in ring and 10 trilithon) encode a 360-day year that will be disrupted (Stonehenge II's inclusion of 60 unconnected bluestones) after a period of *2400 years* (produced from 40 sarcens x 60 bluestones = 2400). Thus, after 2400 years the outer sarcen ring's 360-day year WILL BE NO MORE. This assessment is proven from fact that 2400 years after 3113 BC is 713 BC, when the Sun retrograded 10 degrees and a cataclysm resulted with the termination of the 360-day year to the VAGUE YEAR OF THE MAYA of 365.24 days. These 2400 years was the conclusion of 6 Baktuns (144,000 x 6 days), which is 864,000 days, *864* being the Foundation of Time and in this context means that the *foundation of the year was broken*, and there are *two* breaks in the perimeter embankment.

The alternation in the orbital year from 360 days to 365.24 is why the Mayan Long-Count does NOT END IN 2012 AD, as popularly propagated, but the addition of 5.24 days a year from 713 BC to the end of the Mayan Calendar's total 1,872,000 days (13 baktun), ends in *2046 AD* when NIBIRU passes Earth and the United States is destroyed by a comet-asteroid collision.

Stonehenge II is a lithic model of the mechanics of terrestrial motion and distance from the Sun, demonstrating that in 713 BC and 2046 AD, THE ORBIT OF THE EARTH IS ALTERED.

Much has been theorized concerning the meaning of the Anunnaki Necklace, or necklace of ARURU in Sumero-Babylonian texts. Following the lead of Zechariah Sitchin, many authors have tried in vain to demonstrate that this refers to the Asteroid Belt lying between the orbits of Mars and Jupiter. And while this may have some truth to it, the provable (and verified by crop circles) identity of the Anunnaki Necklace is the Bluestone Horseshoe inside the trilithons of Stonehenge.

The Sumerian traditions claim that the Anunnaki were represented as *blue stones* on the necklace of a goddess, a tradition reflected in the ancient Americas where we find these stones in the form of lapis lazuli and turquoise. As all sarcens and bluestones identify modes of time-keeping in calendars and chronologies that count down toward the return of the Chief Cornerstone (symbolized in the center of Stonehenge as the Altar Stone, just as we find in the center of the Aztec Stone of the Fifth Sun), the bluestone horseshoe represents a final calendrical countdown to the End Times reign of the most powerful of the Anunnaki: the Seven Kings.

This final calendar can be none other than the Anno Domini calendar. The 19 bluestones are an enigma, 19 is a *lunar* number, however in this instance it is *not* the Earth's Moon being referenced here. The bluestone horseshoe represents the former moon of NIBIRU, the Dark Satellite, and was placed *within* the trilithon geometry that forms a pentagram. This geometrical form identifies with the 108-degree angle, with the five-pointed star forming *ten* 108-degree angles for the total sum of *1080*. In mystic traditions 1080 is a lunar number and within this architectural complex we are specifically confronted with the two sums of 19 and 1080, as they are attached to a final calendar. Thus, the 19 bluestones times *108* equal the Anno Domini year of *2052 AD*. . . the return of the Dark Satellite, the *lost moon* of NIBIRU. This date is encoded many times in the Great Pyramid's timeline, as found in *Chronotecture*. This date and concept is further proven by the decoded messages of crop circles, as we will see. For the critic thinking that this is an arbitrarily contrived date because we selected the Anno Domini system (which is the final Calendar), then consider the following.

Stonehenge III was conducted in 1687 BC after a global series of earthquakes destroyed cities all around the world. In 1687 BC the *Sun darkened* when planet Phoenix passed between the Sun and Earth, these strange transits the subject of *When the Sun Darkens*. After studying the Stonehenge cite, Sir Norman Lockyer concluded that the complex was erected in 1680 BC, a pretty close approximate. But this was merely a *renovation* made by Semitic people in the days of the life of Jacob. Unlike Stonehenge I and II, the III site was not a phase of construction, but re-erecting, for after the Great Flood of 2239 BC and the quakes of 1687 BC, Stonehenge had been damaged. According to the bamboo books of the Chinese found in 279 BC, the year 1687 BC was a disaster with meteoric rain when a *five-planet alignment* occurred (see *Chronicon*).

The remarkable fact about 1687 BC is that from the introduction of the bluestones (that symbolize the Anunnaki influence over mankind) into the Stonehenge complex in 3113 BC passed 1426 years, or 713 + 713 years. The 3113 BC Mayan Long-Count's baktun count of perfect 360-day years for a total of *2400* years ended in *713 BC* when the Dark Satellite, an Anunnaki *prison*, nearly collided into Earth. The bluestones, as we have seen, indicate a 2400-year period as well, which ended in 713 BC. These 713 years count down to the Anno Domini calendar, and the 19 bluestones exhibit the release of the Anunnaki from the Dark Satellite in 2052 AD (108 x 19).

### Signs in the Earth

The recent Crop Circle phenomenon is another enigma laden with researchers, for the most part looking in all the wrong directions. They are simple in their mathematical interpretations, despite the complex geometrical designs they appear in. The exact same formulae employed in the science of chronometry, as used in the previous Archive to decode Stonehenge I, remains the same for comprehending the latent messages in these odd and badly misunderstood agriglyphs. There are two facts that need be taken into consideration relevant to our thesis, both facts made known in Andrew Collins' pioneering work entitled *Crop Circles: Signs of Contact*.

First, Collins asserts that about 95% of all crop circles appearing worldwide manifest within *40 miles of Stonehenge*.(2)  Second, researchers have noted that the first crop circles, the first circles appearing with rings and the first double-circles connected with lines were all of the same geometry, scale and proportions as the *Bluestone Horseshoe* of 19 stones within Stonehenge.(3)  This author interprets this fact in one way: in their campaign to delude the governments of the world that they are an advanced extraterrestrial species, the Anunnaki initially began forming crop circles. Some men bought into the deception while others remained skeptical. Because the Anunnaki initiated this form of communication, Watchers still loyal to the Creator were assigned the task of providing messages to mankind in the same way, elaborate *warnings* with absolutely *beautiful* geometric patterns.

Such a thesis of course requires substantiation. Review herein the *signs* given to mankind by the mind of God.

### Stonehenge "Julia Set" (first reported 7-7-1996)

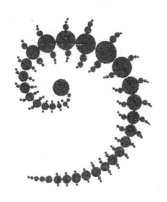

The Julia Set seen in proximity to Stonehenge (1996).

Appeared in close proximity to STONEHENGE on Salisbury Plain in Wiltshire, England in wheat field across the highway in daytime. 900 by 500 feet.

151 total circles in formation; 36-segment formation divided into 115 satellites stemming from segments.

FORMULA:  151-36 = 115. . . 115 x 36 = *4140*

The geometrical crop formation identifies the calendrical function OF STONEHENGE as being a chronolithic CURSED EARTH calendar for 4140, which is exactly ten cursed Earth periods of *414* years each.

The Julia Set is a geometrical phenomenon, a perfect representation of the Fibonacci Spiral. Leonardo Pisano, known to us as Fibonacci, was a guru is medieval western mathematics who traveled widely and published his *Liber Abaci* in 1202 AD. The Fibonacci series concerns a sum that is the total of the two previous numbers in a sequence, a formula that encodes many natures of mystery concerning the harmonics of light and music. Interestingly, the 12th Fibonacci number is 144 (12 x 12). The inventor John Keely, a man way before his time, wrote that this Spiro-Vortex is the key to nature and comprehending the mechanics of the universe through *whole numbers*, which are the only numbers used in chronometry.

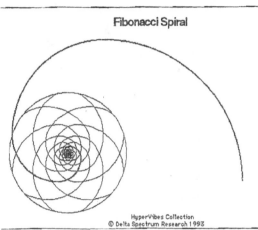

Fibonacci Spiral

HyperVibes Collection
© Delta Spectrum Research 1993

**<u>Triple Julia Set</u>** (first reported 7-7-1996)

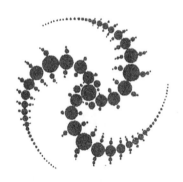

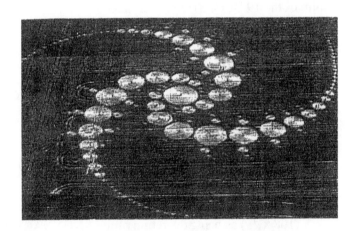

Appeared same day as Stonehenge Julia Set, also in Wiltshire but located at Windmill Hill (ancient megalithic site) near Yatesbury. Measures 260 x 260 meters.

199 total circles
198 circles when eliminating axial circle (only circle not in triplicate)
3 arms of formation have 26 circle segments each (78 total segments)
3 segmented arms have 40 satellite circles each (120 total satellites)

<u>FORMULA</u>:  198-78 = 120/ 120 x120 (satellites) = *14,400* (144 x 100)

The number 120 is encoded twice in this formation and is also found prominently multiple times in STONEHENGE chronometry, being the FOUNDATION number known in antiquity as the Sumerian shar.

The 14,400 sum is too large to refer to years within recorded history, but instead denotes *days*, for this is exactly *40 years* on the Sacred Year count of 360 days a year (360 x 40 = 14,400).

These 40 years is a *countdown* from 1996 (when this appeared) to the year 2036 AD, the 40th year of the Antichrist, who was born in 1996 AD!  2036 AD begins the APOCALYPSE.

## The Pointer Formation (1996 AD)

This formation represents a NECKLACE, specifically the Sumerian goddess ARURU's necklace, each segment being one of her *ANUNNAKI* offspring.

"Pointer" is a misnomer, this agriglyph represents a *comet*, which is verified by its mathematical interpretation. Liddington Castle, Wiltshire.

40 total agriglyphs
20 segments
19 satellites
(13 primary satellites)
(6 tiny outer satellites)
1 "pointer"

FORMULA:  40 - 20 = 20. . . 20 x 20 = 400 (Mayan baktun of 144,000 days)
400 x 13 (primary satellites) = *5200* (13 baktuns of Mayan Long-Count

The NECKLACE with 19 satellites mimics the 19 BLUESTONES of inner Stonehenge, both mathematically and geometrically.

*Hand-drawn diagram labeled:* 19 BLUESTONES; 108° SUMERIAN ARUB IS COMET SYMBOL

* 19 bluestones and 19 NECKLACE satellites form semi-circle horseshoe.

* 19 bluestones surround Trilithon geometry of a *pentagram* formed of 108-degree angles forming Sumerian AR UB *comet* symbol (akin to goddess *AR*-URU) and 19 satellites of NECKLACE surround a *comet* agriglyph.

* Both Stonehenge's bluestones and this crop formation indicate the calendrical year of 2052 AD (108 x *19*) as being the return of the ANUNNAKI. These beings are said to return in the Book of Revelation as the SEVEN KINGS, ruled over by an *EIGHTH*. This 8th Anunnaki King is found in the ancient Sumerian King-Lists and indicated in this crop formation by the fact that the NECKLACE identifies the Anunnaki and also, the *8th* Anunnaki circle is "pointed" at by the tail of the *comet*. According to Revelation, it is by *cometary impact* that the "locusts" (Anunnaki) emerge on Earth in the Last Days.

## Fractal Formation (2005 AD)

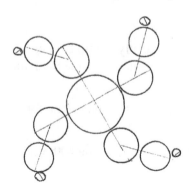

Appeared in Poland

13 total crop circles

FORMULA:  13 x 4 = *52*

4 arms

Geometrical Interpretation:  8 angles all at 135 degrees is *1080* (135 x 8)

## Cuxton Formation (1998 AD)

27 total circles

4 unusually large equi-dimensional circles

FORMULA:  27 x 4 = *108*

This formation appeared in 1998 AD, which is exactly a *108-year* countdown to the Year 6000 Annus Mundi (2106 AD), the year the Anunnaki will suffer defeat under the dominion of the Chief Cornerstone at Armageddon.

## Tri-Armed Triskele Formation (1997 AD Cuxton)

21 total circles
12 ring formation
9 circles in three arms

FORMULA:          21 - 12 = 9
    9 x 12 = *108*

FORMULA:          360-degree ring formation
    360 x 3 (arms) is *1080*

## Ogbourne St. George Formation (2003 AD)

18 circles (actually 19, but axis omitted)
6 point formation

FORMULA:  18 x 6 = *108*

## Milk Hill Formation (2001 AD)

409 total crop circles

408 (omitting axis circle, which is the only one not in hexagrammic formation)

6 segmented arms of 13 circles each (78 total segments) – 78 is the number of segments in *Triple-Julia Set*

330 satellites (55 satellites on each of the 6 arms)

FORMULA A:  408 circles x 6-fold pattern is *2448*. This is the ancient Annus Mundi year of 1447 BC, the *Exodus* of the 13 Tribes of Israel from Egypt (historic type of EVIL WORLD), when Israel was led into the *LAND OF PROMISE.*

FORMULA B:  (Same formula as Julia Sets!)

408 - 78 = 330. . . 330 x 78 = *25,740*. Just as in *Triple Julia Set*, this 25,740 is too large to be years in recorded history, but instead represents *days*, a *countdown* from the appearance of this crop formation of 25,740 days to a future event.

25,740 days identifies the year *2070 AD*. 69 years is 25,185 days (365 x 69), but we must add 17 days for leap years for 25,202 days. Differential from 25,740 of crop formation and 25,202 in 69 years is 538 days. But this crop formation did not occur in December of 2001 but earlier in the year, thus these 538 days is eliminated from the months in 2001 to year's end and the exact time in 2070 AD that the RESURRECTION occurs.

This formation mathematically refers back to the Exodus but calendrically points to the future *PLANETARY EXODUS* of God's chosen ones from apocalyptic earth (foreshadowed in 10 Plagues on Egypt). 2070 AD is equal to 414 x 5.

### Uberdingen, Baden-Wurttemberg, Germany (2006 AD)

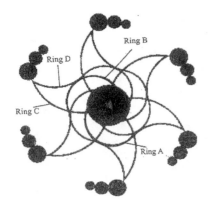

18 crop circles
6 arms

18 x 6 = *108*

**Nine-Pointed Star Formation:** Cherhill (1999 AD)

9 triangular points
15 total is sum of Stonehenge trilithons
6 crescent arcs

The 9 triangular points have 2 faces each (18)
18 x 6 arcs = *108*. . . 108 degrees found conceptually in Stonehenge's Trilithon geometry.

**Wilton Windmill Formation** (2004 AD)

12 circles on perimeter
12 pyramids (6 facing inward / 6 forming inner star)
6-point formation

FORMULA:  12 x 12 = 144 . . . 144 x 6 = *864* (Foundation of Time number)

Note: Star of David seen inside formation is the geometric symbol for number *108* (6 pyramids have 18 faces: 6 x 18 = 108)

**Beckhampton Formation** (2001 AD)

96 pyramids
24 other forms          96 + 24 is *120* (Foundation number)

Outer ring has 8 pentagrammic Anunnaki NER symbols (identifying 108 geometry). 108 x 8 = *864*.

Anunnaki NER is 600 years, or 5 (pentagram) periods of *120 years* (120 x 5 = 600). The sum of 120 is double-referenced, for the outer ring of triangular pentagrams geometrically displays 8 pentagrams made of 5 three-sided triangles, each pentagram having *15 sides* (again reflecting 15 Trilithons of Stonehenge). . . 15 x 8 = *120*.

**Milkhill, Wiltshire Formation** (1997 AD)

 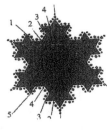

18 pyramidal perimeter points (48 facets)
126 crop circle satellites
126 - 18 = *108*

126 satellites are:

54 large circles (½ of 108)
72 tiny circles (½ of 144)

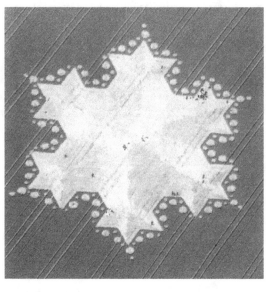

48 perimeter facets x 18 points = *864*
48 x 54 large circles = 2592 (*864* x 3)
48 x 72 tiny circles = 3456 (*864* x 4)
48 x 126 total satellites = 6048 (*864* x 7)
864 is Foundation of Time

Two principle sums of this formation are *108* and *864*. . .
and 108 x 8 = *864*.

### The Aldbourne Formation (2005 AD)

*408* total is same at the *MILK-HILL Formation* of 2001 AD that encoded the date of the RESURRECTION and historic EXODUS. This same 408 x 6 (6 is layers of geometric form) is 2448 Annus Mundi, year of Exodus (1447 BC).

1    ring-circle perimeter

2    30 four-sided forms of outer ring of diamonds (30 x 4 = *120*)

3    27 four-sided forms of next ring of diamonds (27 x 4 = *108*)

4    21 four-sided forms of next ring of diamonds (21 x 5 = *84*)

5    15 four-sided forms of next ring of diamonds (15 x 4 = *60*)

6    9 four-sided forms of inner ring of diamonds (9 x 4 = *36*)

The 6 layers add up to 2448, but the central icon is a 3-sided triangle. Thus, 2448 x 3 is *7344 years*, the ENTIRE LENGTH OF THE GREAT PYRAMID'S GEOMETRICAL CALENDAR from 5239 BC to 2106 AD, the year 6000 Annus Mundi. By 2448 AM (1447 BC), mankind had survived under Anunnaki influence for a *third* of Anunnaki history, and the PLANETARY EXODUS of the Resurrection in 2070 is when the Redeemed among men have filled the positions of the fallen Anunnaki, the *third* of heaven that rebelled.

### Double Helix Formation: England (1996 AD, June)

12 circles
78 satellites forming double-helix
(78 is sum of segments in Triple-Julia Set of *1996* and sum of segments in Milkhill formation of 2001)

FORMULA:     78 - 12 = 66 (abbreviated 666)
             66 x 12 = *792* (orbital duration of Anunnaki planet NIBIRU)

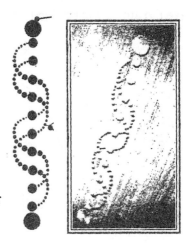

MESSAGE:  The Anunnaki have perfected their angel-human hybrid DNA pattern and executed it in the form of a man born in *1996* who will become the ANTICHRIST, their terrestrial representative who will deceive mankind into believing he is God returned in the form of a man (Messiah). This DNA-formation is 34 years after the 1962 AD end of the *Anunnaki NER Chronology* (144 yrs. to Armageddon), 34 years being 408 months. This formation appears to be attached mathematically, geometrically and conceptually with the following other crop formations:

1.     Stonehenge Julia Set of 1996

2.     Triple-Julia Set of 1996

3.     Pointer Formation of 1996

4.     Milk-Hill Formation of 2001

5.     Aldbourne Formation of 2005

## Avebury Manor, Wiltshire Formation (1999 AD)

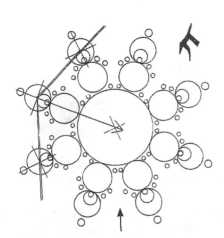

11 crop circles forming each arm
8 arms for total of 88 crop circles (omitting axial circle)
48 tiny circles are satellites

FORMULA:  88 - 48 = *40*. . . 40 x 48 = *1920*.

1920 AD is the year 5814 Annus Mundi, which begins a 186 year countdown to the year 6000 AM (2106 AD) – Armageddon and return of the Chief Cornerstone who defeats the Anunnaki, this being significant because 5814 is the precise Pyramid Inch height of the Great Pyramid, once fitted with its Apex Stone (monument currently has no cornerstone).

This interpretation is corroborated by fact that this crop formation is octagonal in dimensions, this 8-sided glyph having exactly 8 angles of 135 degrees. 135 x 8 = *1080*, this being the *Annus Mundi year* of 2815 BC, when the Great Pyramid was finished and left without a cornerstone (who would not arrive, as an actual person, until the Year 6000 AM: 2106 AD).

The odd symbol in upper right of formation is unknown to author.

### Five Pyramids Formation (1999 AD)

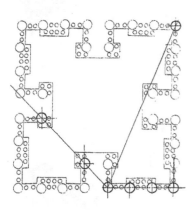

Located at Long Barrow, West Kennet, Wiltshire.

156 satellites
36 large satellites
120 tiny satellites

FORMULA: 156 - 36 = *120*. . . 120 x 36 = *4320* (1080 x 4 or 1440 x 3)

Variant Formula: Large "five pyramids" glyph has 68 facets. . . 68 x 36 large satellites is *2448*. . . the Annus Mundi date of the Exodus in 1447 BC. Further, 120 - 68 = *52*.

Variant Formula: 156 satellites – 68 facets is *88*. . . 88 x 68 is *5984*, the Annus Mundi Year of *2090 AD*, the final and worst of the SEVEN PLAGUES of Revelation. Thus, 2448 identifies the 10 Plagues on Egypt and 5984 is the end of the PLAGUES of the Apocalypse in 2090 AD.

### Fyfield Down Formation, Wiltshire (1999 AD)

37 principle circles (36 omitting axial circle not in hexagrammic form)
54 satellites
84 circles total
36 outer perimeter facets
84 + 36 = *120*

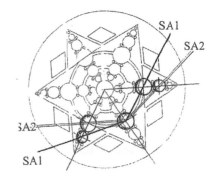

FORMULA:    120 x 12 point terminations is *1440* (144 x 10)
1440 x 6 (6-pointed form principle design) is *8640* (864 x 10)

### Hackpen Hill, Wiltshire (1997 AD)

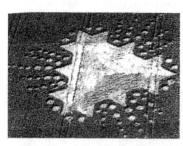

Crop formation is composed of 13 pyramids: 12 perimeter pyramids on a large pyramid.

33 large circles
62 tiny circles

FORMULA: 33 x 62 = *2046*

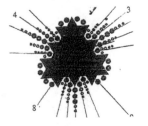

Tiny and large circles are configured to appear as *approaching comets* and these identify the calendrical year of *2046 AD* when the United States will be destroyed by cometary impact, bringing to an end the global reign of the *13th Tribe of Israel* (USA and Britain), also identified here as the large pyramid bordered with the 12 perimeter pyramids (12 Tribes of Israel).

Verification of Interpretation:

This crop formation appeared in 1997. This appearance happens to be an ISOMETRIC PROJECTION verifying the interpretation of this formation's meaning of the 2046 AD end of the United States (13th Tribe). 2046 AD is exactly *49 years* (7 x 7) after the 1997 AD appearance of this crop glyph, and *49 years* before 1997 AD is the year *1948 AD* REBIRTH OF THE STATE OF ISRAEL, giving the internationally dispersed people of the 12 Tribes a place to live in the land of their forefathers.

**Winchester, Hampshire Formation** (June 22nd, 1995)

Discovered at Longwood Warren near Winchester was this unique depiction of our Solar System depicting the Sun, Mercury, Venus and Mars, as well as the Asteroid Belt. Earth itself is *missing*, and to confirm its absence from its position, its orbital belt is clearly shown. Astronomers Gerald Hawkins and Robert Hadley both confirmed that this was an accurate model of the solar system. The cellular changes in the plants prove this crop circle to be original and not a hoax.(4) The key to interpreting the time when Earth will be removed from its place is *time* itself. The Asteroid Belt consists of 52 clockwise circles and 13 counterclockwise circles,(5) these numbers being the denominators for the Mayan Long-Count that ends in 2046 AD. The 2046 AD event of NIBIRU's return is 51 years after the appearance of this sign, and the Isometric year from 1995 is 51 years prior, or 1944 AD. . . the year the USA landed the most expansive *amphibious invasion* upon a foreign shore (Normandy) in world history in Operation Overlord, mirroring the Anunnaki invasion that will occur in 2046 AD, when Earth is moved from its place by the arrival of NIBIRU.

We have reviewed 20 *signs* from the mind of the Creator specifically designed to warn His creations, humanity, of the designs of the Anunnaki and the timeline these baleful entities are subject to. The conspiracy theories concerning men in black as being government operatives is not accurate – they are actual agents of the Anunnaki sent to deter investigators, intimidate them and lead them astray, as they do to the government agencies around the world that unofficially study these signs. The occasionally witnessed black helicopters associated with these crop circles do not manufacture them, but photograph the site for insiders in the government who seek to interpret these agriglyphs. Most of the officials in world governments who are studying the phenomena have been duped by Anunnaki agents into believing that these signs are attempts to communicate by an advanced alien race.

But the truth is that this "alien race" is not alien to this world. . . they owned the planet before we did. And they will be returning to rule over the world, which was, in the beginning, rightfully theirs. Those who have become attached to this life and the world in its present state will learn the error of their choice in 2046 AD. Those of us who await the return of the Chief Cornerstone await another life, in another time and place, a destiny that by rebellion the Anunnaki lost.

*Archive 14*

# Noahic Warning Comet and the
# Spear of NIBIRU Apocalypse Comet


The Egyptians keep written records of the observed results of any
unusual phenomena, so they come to expect a similar consequence to
follow a similar occurrence in the future.

—Herodotus, *The Histories* (1)


The comet groups previously detailed in this work were from antiquity and sufficient enough to introduce our profound discoveries concerning 2046 AD and the year 2052 AD. These last few Archives are found at the end of this book because inclusion earlier in the text would have unnecessarily broken up the continuity of the research, and further, the rest of these orbital timelines concern only astronomical events occurring after 1260 AD. These groups were all brought into their own orbits after NIBIRU made its final pass into our inner solar system, entering in 1254 AD and exiting in 1314 AD, causing terrible disasters around the world. As will be seen, we are not entirely finished with 2046 AD.

In 1694 AD astronomer Edmund Halley proposed before the prestigious Royal Society of London that the Great Flood had been caused by a *comet collision*.(2)  He studied European records for the last few centuries and determined that a comet would pass through the inner system and be seen clearly from Earth in 1758 AD. This comet did indeed appear, after his death, but his name was attached to it and it has since been known as Halley's Comet.

For reasons that will become clear, this author calls it the Noahic Warning Comet. Allegations that Halley's Comet was seen prior to 1260 AD are incorrect; comets prior to this date are of other groups already detailed, nor do they fit the orbital pattern here. Astronomers often use approximates to cover errors and further conceal their mistakes by claiming comets or other orbiting objects are *perturbed* by gravitational forces that slow them down or speed them up, whichever fits their theoretical model at the time.


**Noahic Warning Comet**

After breaking away from NIBIRU, this solitary comet (not a part of a group) passes through the inner system and over the ecliptic in 1306 AD. Orbit varies from 75.3 to 75.7 years.

| | |
|---|---|
| 1306 AD<br>(5200 AM) | This is during the chaotic period known as the Seven Comets, a period of 16 years from 1298-1314 AD when catastrophes were recorded around the world. This is the 5200nd year of the Annus Mundi chronology, paralleling the Mayan Long-Count's 144,000 x 13 days, or 1,872,000 days of 360 years. This date is also noted for the decline of the Mayan civilization. This year begins an 800-year countdown to Armageddon and the return of the Chief Cornerstone in 2106 AD (6000 AM). |
| 1381 AD<br>(5274 AM) | No records. |
| 1456 AD<br>(5350 AM) | Antonion Bonfini and other Europeans left records of an unusual comet with *two tails* and half as long as the sky with golden-like flames. |
| 1531 AD<br>(5425 AM) | Halley's Comet witnessed by the prophet Nostradamus.(3) |
| 1607 AD<br>(5501 AM) | No records; this is the year the first permanent English colony was founded in North America – Jamestown. |
| 1683 AD<br>(5577 AM) | No records. |
| 1758 AD<br>(5652 AM) | Comet seen and named Halley's. Though this author found no records for the prior 1607 and 1683 passings, Halley must have had them in order to make this prediction. |
| 1834 AD<br>(5728 AM) | No record of a comet, but astronomer Pastorff reported that large fragments, huge in size, passed over the disk of the Sun.(4) |
| 1910 AD<br>(5804 AM) | Halley's Comet is photographed for the first time. It is very visible and exhibited *five tails* (the Exodus 1447 BC comet had ten tails that enveloped earth).(5) |

1986 AD
(5880 AM)

Halley's Comet passed through the inner system and the European spacecraft *Giotto* measured its tail at 10 million miles in length and took the first ever close-up pictures of a comet's nucleus (its *foundation*). Before the Flood, old Noah was warned that the *foundations* of the world would be broken and the world would be destroyed by water in 120 years, the Hebrew word *foundation* having a geometrical value of 120.(6) This year of 1986 AD was precisely *120 years* before the year 6000 Annus Mundi (2106 AD) when the Chief Cornerstone returns upon the foundation of the Monument of man to initiate the Stone Kingdom (Millennium), this Kingdom's foundation being the 6000 years of humanity's trial and testing period. Christ is the Stone the Builders rejected, foreshadowed also in 1986 AD by the pass of the five-mile wide asteroid Adonis (Syro-Phoenician for *Lord*) 186,000 miles from Earth, or 52,000 miles closer than our own Moon!(7) In 1986 AD a major discovery concerning the foundation of Earth was made by Robert Gentry, who found that billions upon billions of polonium 218 radio haloes are entombed within granite around the world. Granite is basement rock, the foundation or skeleton of the planet. It requires about three minutes for radio haloes to appear, but they quickly vanish. The mystery is how they appeared in solid granite, for the GRANITE MUST HAVE ALREADY BEEN HARD WHEN THE HALOES APPEARED, for the haloes would have long since dissipated during the cooling process. Creating radio haloes in granite is *impossible*, for molten granite turns to ryolite. Thus the halo's presence in granite means that the GRANITE WAS FORMED IN AN *INSTANT*.(8) Granite, with its very large crystals, cannot be made from molten rock and the radio haloes captured in the stone are evidence of INSTANT CREATION that occurred at the FOUNDATION of the world, made known to man in this 120th year before Armageddon. It is also to be noted that in 1986 AD crop circle researchers discovered the first crop formation appearing with *rings* around the agriglyphs.(9)

This is the final pass of Halley's Comet important to our thesis, for the next is scheduled in 2060 AD, however, the 2046 AD orbital alteration of Earth will change this. This comet may have a future, darkly apocalyptic role to play in the tribulation drama that will occur.

### Spear of NIBIRU Apocalypse Comet Group

This train of asteroids, ice-encased debris and a large comet broke free of NIBIRU when the Anunnaki Homeworld entered the system in 1254 AD. The strewn train is 7 to 8 years long, initially being only 6 years long.

1254-1260 AD
(5148-5154 AM)

Initiate their break from NIBIRU and begin solar orbit.

(61.9 years over ecliptic)

| | |
|---|---|
| 1315-1321 AD<br>(5209-5215 AM) | In 1315 AD bad harvests were reported throughout Europe, resulting in a famine.(10)  Famine lasts through 1517 and results in incidents of cannibalism. It has been estimated that 10% of Europe's population died. |

(18.5 years under ecliptic)

| | |
|---|---|
| 1333-1339 AD<br>(5227-5232 AM) | In 1333 AD a meteorite crashed into China killing everyone within 100 miles with noxious gases.(11)  Strange omens are seen in the skies over China and Mongol Emperor Toghon Temur Khan ascends the throne. Many Asian records attest that this time was one of earthquake activity, weird storms and comets.(12)  In 1336 AD extraordinary thunderstorms are reported over France. |

(61 years over ecliptic)

| | |
|---|---|
| 1394-1400 AD<br>(5188-5294 AM) | In 1400 AD a meteorite of metal crashed into a region close to Elbogen, Bohemia. When it was found it still weighed 235 lbs.(13) |

(18 years under ecliptic)

| | |
|---|---|
| 1412-1419 ADM<br>(5306-5313 AM) | No records. |

(61 years over ecliptic)

| | |
|---|---|
| 1473-1480 AD<br>(5367-5374 AM) | 1480 AD was a very important pyramid chronometrical year. Arabian records tell of a comet that appeared like a spear with a sickle in 1479. (14) |

(18 years under ecliptic)

| | |
|---|---|
| 1490-1498 AD<br>(5384-5392 AM) | In 1490 a meteorite shower in central Asia killed 10,000 or more people (15) and on the other side of the world a volcano erupted in the Ox Mountains of Ireland, killing many people and much livestock.(16) |

(61 years over ecliptic)

| | |
|---|---|
| 1551-1559 AD<br>(5445-5453 AM) | In 1551 AD a strange epidemic called Sweating Sickness afflicted England. This disease has never resurfaced since this date, and its prior four appearances (see *Chronicon*) seem to be associated with astronomical phenomena. In 1556 AD on January 23 more than 800,000 people perished in an earthquake in the Shensi Province of China, principally in Shaanxi. |

(18 years under ecliptic)

| | |
|---|---|
| 1569-1577 AD<br>(5463-5471 AM) | In 1577 AD a bright comet appeared over Europe.(17)<br>In October of 1571 a bright streak appeared in the night sky for a long time, as seen from Europe. It was a huge pillar of fire regarded as a good omen to the combined naval fleets of Spain, Venice and the Roman papal States on their way to battle the Turkish Ottoman navy. On October 7th the Catholic fleets defeated the Turks in the naval Battle of Lepanto.(18) In 1577 in Altorff near Wittenberg, Germany, was seen black objects in the sky in December.(19) |

(61 years over ecliptic)

| | |
|---|---|
| 1630-1638 AD<br>(5524-5532 AM) | In 1631 AD Mount Vesuvius erupted in Italy killing 4000 people, far less than it had suffocated in 79 AD and previously in 1687 BC. Otherwise no records. |

(18 years under ecliptic)

| | |
|---|---|
| 1648-1656 AD<br>(5542-5550 AM) | R.P. Greg reported luminous objects falling out of the sky in 1652 AD.(20)  In Chladni's account, a viscous mass fell with a luminous meteorite between Rome and Siena.(21)  This same year concludes the 12th baktun of the *Mayan Long-Count* (144,000 x 12 days). |

(61 years over ecliptic)

| | |
|---|---|
| 1709-1717 AD<br>(5603-5611 AM) | The train seems to have spread into a ninth year by this date. In 1718 AD the exact same phenomenon recorded in 1652 AD happened again. Upon the island of Lethy, India, rained globes of luminous objects, meteorites, that when investigated turned out to be some gelatinous matter, as recorded in the American Journal of Science.(22) |

(18 years under ecliptic)

| | |
|---|---|
| 1727-1736 AD<br>(5621-5630 AM) | This period coincides perfectly with the passing of the 2046 AD NIBIRU Comet Orbit group that moved through the inner system from 1730-1736 AD and involved 5 continuous years of earthquake activity, quakes attended by luminous clouds moving rapidly across the sky, strange mists in the sky that glowed, influenza outbreak in the Americas and a rain of dust and soil, all described earlier in this book. |

(61 years over ecliptic)

| | |
|---|---|
| 1788-1797 AD<br>(5682-5691 AM) | In 1788 AD astronomer Schroeter witnessed a light appearing on the Moon in the lunar mountains. But when sunlight later illuminated the region, he saw that it was cast in a great blackness.(23)  This appears to have been the eruption of a lunar volcano; eruptions were first noticed the year previously, in 1787 AD, when Herschel first reported these lights he believed were eruptions.(24)  In 1792 AD Mount Unzen in Japan erupted killing 14,500 people. This date is 1000 + *792* Anno Domini. In 1794 AD in the month of June, Mount Vesuvius erupted and stones rained from the sky over Tuscany.(25)  Also, during a violent storm, stones rained over Siena, Italy.(26)  In 1796 AD we have the return of the glowing meteorites of gelatinous matter falling at Lusatia in March.(27)  In 1797 AD an earthquake at Quito, Ecuador killed 41,000 people. The following year in 1798 AD it rained a strange bituminous substance over Germany in March.(28) |

(18 years under ecliptic)

| | |
|---|---|
| 1806-1815 AD<br>(5700-5709 AM) | In 1806 AD on March 15th rained a bituminous substance near Alais, France. A commission appointed by the French Academy determined it to be like coal.(29)  On May 16th, 1808 AD, occurred the following: |

> ". . .the Sun suddenly turned dull brick-red. At the same time there appeared, upon the western horizon, a great number of round bodies, dark brown, and seemingly the size of a hat crown. They passed overhead and disappeared in the eastern horizon. Tremendous procession. *It lasted two hours*. Occasionally one fell to the ground. When the place of the fall was examined, there was found a film, which soon dried and vanished. Often, when approaching the Sun, these bodies seemed to link together, or were then seen to be linked together. . . and under the Sun, they were seen to have tails three or four fathoms long. Away from the Sun these tails were invisible. Whatever their substance may have been, it is described as gelatinous – 'soupy and jellied.'"(30)

In July of 1811 a meteorite fell from the sky after an aerial explosion. It was investigated where it fell near Heidelburg, where it was discovered to be of gelatinous composition.(31)  A massive earthquake shook the eastern United States along the Mississippi River Valley and further east. In 1812 AD, January, another earthquake shocked the eastern United States and a third more destructive quake happened in February covering 1.5 million square miles. In Russia, a typhus epidemic broke out, crippling the Russian military and afflicting Napoleon's forces as well. A red, viscous matter rained at Ulm.(32)  In March, 1813, a yellow powder rained on Gerace, Calabria which, when examined, was found to have traces of resin. This unusual phenomenon was attended with loud explosions in the sky.(33)  The train was followed in 1816 AD by a huge object, seen as an "extraordinary spectacle," passing through the sky that apparently missed the Earth.(34)

(61 years over ecliptic)

1867-1876 AD
(5761-5770 AM)

In 1867 some carbonaceous matter fell at Goalpara, India, about 8% of it hydrocarbons.(35)  In 1868 on January 30th, a mass of burning sulfur fell from the sky upon a road and was stamped out by bystanders and more carbonaceous matter fell on Ornans, France on July 11th.(36)  On February 24th during a violent storm in 1868 over Ontario there fell with the snow a dark-colored alien substance that, when investigated, was discovered to be a vegetable matter "far advanced in decomposition." This material fell upon a frozen earth in North America in a 50 mile strip 10 miles wide and was estimated at 500 tons.(37)  In 1870 on February 12 a large luminous object was seen in the sky over Italy when an earthquake occurred, followed the next day on the 13th with a rain of *sand*.(38)  On the 14th of February there fell from the sky at Genoa, Italy a yellow substance containing, as seen under a microscope, globules of cobalt blue. In April and May, reports asserted that a lot of this matter fell upon France and Spain, that it was certainly not pollen and that it gave off an odor of charred animal matter.(39)  On August 20th over Switzerland, large salt crystals rained along with hail.(40)  In 1871 the Great Fire of Chicago occurred, aggravated by gale-force winds that created an incendiary tornado so hot that it even burned granite, marble and iron.(41)  A huge fiery ball was seen flying through the sky just prior to the onset of the fire, and other areas around the city in the country and surrounding counties caught fire as well. Astronomers Hind and Denning in November, 1871 report an object seen passing over the Sun.(42)  In 1872 on March 9th, 10th and 11th a dust-like substance fell from the sky reported to have been composed of red iron ochre, carbonate of lime and organic matter.(43)  In November a brilliant meteor shower exhibited over 150,000 shooting stars within a six-hour period. Astronomers, at a loss to explain the unpredicted phenomena, said that these were merely the fragments of comet Biela that split apart in 1845. This Anno Domini year 1872 is an abbreviated Mayan Long-Count of 144 x 13 years. (1872)  In January, 1873, burning cinders fell from the sky and landed on the deck of a ship, and at Marlsford, England on September 4th there fell a strange black rain, and then, 24 hours later, another blackish rain fell again at the same location.(44)  On March 14th, 1873 there rained orange hailstones on Tuscany and in the following year, on June 9th, there fell a hail containing carbonate of soda.(45)  Also, in 1874 AD during the transit of Venus, there happened an earthquake near Cairo, Egypt.(46)  In July fell hailstones of frozen vegetable matter on Toulouse, France.(47)  On April 14th of 1875 at Victoria, Australia a meteorite landed near a residence. When investigated the next morning the area was covered in *cinders*, and in December the Rio de Janeiro Observatory reported seeing vast numbers of bodies passing over the Sun, some being luminous (clear) and others dark.(48)  On March 3rd, 1876 at Olympian Springs, Kentucky, flakes of a substance that looked like beefy leaves fell from a *clear sky*. It was apparently highly pressurized detritus that covered the trees and field of a 300-foot-long and 150-foot-wide area.(49)  On April 20th during a rain storm, a meteorite of iron fell near Wolverhampton, England.(50)  The following year in 1877 AD, February 27th, a yellow-golden matter fell at Peckloh, Germany, in which four kinds of organisms were detected that were described as microscopic arrows, coffee beans, horns and disks.(51)  On June 23 a reddish rain fell on Italy, and on October 14th, during gale-force winds, something fell from the sky that looked like a bright luminous green fire.(52)

(18 years under ecliptic)

1885-1894 AD
(5779-5788 AM

In 1885 a volcanic island, Falcon Island, emerged from the sea. On August 10th over Grazac, France fell a bituminous matter.(53)  At midnight, February 24th, 1885, latitude 37 north and longitude 170 east, between Yokohama and Victoria, the captain and crew of the Innerwich noticed the sky turning fiery red. A couple minutes passed before, "All at once, a large mass of fire appeared over the vessel, completely blinding the spectators." The fiery mass fell into the sea. Its size may be judged by the volume of water it displaced. The noise was deafening and the sea became tumultuous and almost knocked over the ship.(54)  On November 1 a giant red object moved slowly in the sky and was four times the size of the Moon, as seen from Adrainople.(55)  On August 27th, 1885 at about 8:30 AM there was observed by Mrs. Adeline D. Bassett, "a strange object in the clouds, coming from the north."  She called the attention of Mrs. L. Lowell to it, and they were both somewhat alarmed. "However, they continued to watch the object steadily for some time. It drew nearer. It was of triangular shape. . . . While crossing the land it had appeared to descend, but, as it went out to sea, it ascended, and continued to ascend, until it was lost to sight high in the clouds."(56)  One wonders if this was an Anunnaki vessel. On March 19, 1886 at 3 PM, a profound darkness covered the city of Oshkosh, Wisconsin, in five minutes becoming as black as midnight. This blackness lasted for 8 to 10 minutes on a day that had been light but cloudy. It came out of the west and passed over Oshkosh going east, and reports came from towns west of Oshkosh claiming that it had already occurred there. A wave of total darkness passed from west to east.(57)  During a strong gale on March 1, 1886 a meteorite was seen.(58)  On April 15th of 1887 a resinous substance fell and on April 30th there fell upon Castlecommon, Ireland, a thick blackish rain.(59)  On March 19th, 1887 during a severe storm at sea, as seen from a Dutch bark, two objects were seen in the sky, one luminous and the other dark. One or both crashed into the sea with a roar and large waves. Immediately afterward, chunks of ice fell from the sky.(60)  On March 6,

1885-1894 AD
(5779-5788 AM)

[Cont'd]

1888 a reddish rain fell upon the Mediterranean, and 12 days later it rained a red substance again. When researchers applied heat, the substance gave off a strong odor of animal matter.(61)  On March 9th a block of limestone fell from the sky near Middleburg, Florida.(62) On August 14th a black rain fell so dark it was described as a *shower of ink*, at the Cape of Good Hope. In 1889 an immense quantity of black pebbles rained upon Palestine, Texas.(63)  They were not like any geological formation in the region. On April 2nd, an intense darkness covered Aitkin, Minnesota, followed by a rain of sand and solid chunks of ice.(64)  On June 9th a substance fell in the skies of Russia that yielded forth an unknown mineral.(65)  In June, 1890, hail mixed with sandstone pebbles rained on France, and on May 15th a substance like blood rained upon Messignadi, Calabria. It was sent to Rome for testing and determined to be *real blood*.(66)  Obviously, the residue of an ancient war or disaster when NIBIRU was quick-frozen, its blood-flooded fields later broken off into cometary glaciers awaiting the day they would rain on Earth, entering the atmosphere as a time-capsule of divine wrath. On October 27th, a comet-like body moving at an extreme velocity covered 100 degrees as observed from Grahamstown in 1890, making this distance of 45 minutes.(67) Evidently, it was extremely close to Earth. On January 20th 1891, a meteorite was seen over Italy, followed by a rain of stones and an earthquake.(68)  On April 4th 1892 AD a Dutch astronomer, Muller, witnessed black objects traveling over the face of the *Moon*.(69)  In October 1893, comet Brooks collided into something in space and photographic evidence of this crash was published in the February 1894 edition of *Knowledge*. The collision was with an unknown dark object.(70)  On November 30th during a total eclipse of the Sun by the Moon over Santiago, Chile, tens of thousands of people gazed into the heavens in amazement at a mysterious red nebulous veil enveloping the Earth. After the nebulous matter passed and while the Sun was still eclipsed, a dark object was clearly visible as it passed over the orb of the shadowy Moon.(71)  A similar object was seen by the Lowell Observatory to be near Mars in 1894 AD, on November 25th.(72)

(61 years over the ecliptic)

1946-1955 AD
(5840-5849 AM)

A polio epidemic broke out in the United States. Unfortunately for our thesis, the knowledge filters erected by academia in the early 20th century were by now in full effect, the result being that any unusual phenomena that could not be explained was either ignored or rationalized away by simple explainable occurrences. The majority of astronomers refuse to report truly enigmatic sightings due to attacks against their prestige, character and abilities as professional scientists. On January 15th 1951, Mount Lamington, New Guinea erupted, spewing toxic gases for five days, killing 5000 people. Astronomers reported to the U.S. government that large objects were heading toward Earth from space, thought to be giant asteroids. But at a distance from Earth these objects adopted a geosynchronous orbit around the planet at the equator and their presence as *vessels* (1953) was discovered and kept secret. A dialogue between these Anunnaki visitors and the U.S. officials resulted in an alliance for the exchange of goods (knowledge given to humans, genetic material rights over abducted humans given to Anunnaki Watchers).(73)  By June of 1953, already *250 tornadoes* had afflicted the United States. These Anunnaki vessels, shaped like huge asteroids, arrived with the Spear of NIBIRU Apocalypse Comet group and quickly departed. In 1954 AD, comet 12P/Pons-Brooks appeared and was seen for several months, brightly lit and giving off several unusual explosions between July and September.(74)  Origin of this comet is unknown. Whether it belongs to the Spear of NIBIRU group remains to be seen. On November 29th an 8.5-pound meteorite crashed through the roof of a house of Sylacuaga, Alabama, through the ceiling, struck a radio and bounced before hitting Anne Elizabeth Hodges on the hip, leaving a dark bruise. A famous case.

(18 years below ecliptic)

1964-1973 AD
(5858-5767 AM)

A powerful earthquake afflicted Alaska in 1964 but due to remoteness only 131 people died. A windstorm at Bangladesh, India killed 17,000 people, followed by another storm killing 30,000 people and a *third* storm in December that killed 10,000 more people at Bangladesh. A great blackout occurred in the northeastern United States, leaving millions without power. Power companies not connected to the main power grid were also affected and VLF (Very Low Frequency) and ham radio reception were jammed with static, all indicating not a power outage, but an electromagnetic disturbance.(75)  In 1970 a quake in Peru caused an avalanche in May that buried alive and killed 25,000 people. A total of 66,000 died due to the quake. A violent cyclone killed 300,000 people at Bangladesh, India. In 1972 a large meteor the size of a football field blazed across the sky over North America, witnessed by many in the United States and Canada as a fiery ball in the daytime.(76) In 2009 a comet was seen called McNaught, the largest ever measured. Unknown if it belongs to the Spear group.

(61 years over ecliptic)

| | |
|---|---|
| 2025-2034 AD<br>(5919-5928 AM) | As this date has not occurred, the reader is left to speculate for himself on what will unfold. Or, if a peek into the future is desired, refer to *Chronicon: Timelines of the Ancient Future.* |

(18 years under ecliptic)

| | |
|---|---|
| 2043-2052 AD<br>(5837-5946 AM) | These years are the most devastating for Earth, with the most human casualties, especially in the year 2046 AD. This period ends in 2052. |

It is to be remarked that the majority of these astronomical sightings or events happened in March, April or May of each year or took place within this entire three-month period. We are indebted to the exhausting research of Charles Fort, who was able to penetrate the frontiers of occult science before these priests of knowledge were able to activate fully their knowledge filters.

"An approach is needed that can help break down the compartmentalization of knowledge—in itself a consequence of man's propensity to split apart and artificialize phenomena that are functionally unified."

—Trevor James Constable in
*The Cosmic Pulse of Life* (77)

## Asteroid 2046 AD Impact Orbit Group
## and the Giza Impact Comet

"Time is the number of the motion of the celestial bodies."

—Proclus, *On Motion* (1)

In this Archive we explore the orbital chronologies of the final two groups having appreciable timelines, fixed orbits that despite the increasing entropy of the immense debris fields orbiting the Sun derived from NIBIRU, exhibit through history their reappearance over and again with predictive periodicity. This deteriorating group of asteroids and comets maintains a fixed 92-year solar orbit and is spread in an apparent 4-year train.

1268-1271 AD          In 1268 AD an earthquake at Cilicia in Asia Minor killed 60,000 people. A
(5162-5165 AM)        large asteroid in the group passed over ecliptic in 1270 AD.


      (40 years over ecliptic)


1308-1311 AD          This period is amidst the 16 years of the Seven Comets reported widely
(5202-5205 AM)        over Europe and China, as revealed earlier in this book. Huge asteroid
                      passed in 1310 AD.


      (52 years under ecliptic completing 92 year orbit)


1360-1363 AD          In 1362 AD a Norse-Goth expedition sent by King Magnus of Norway
(5254-5257 AM)        explored North America. They were attacked by Indians in Minnesota and left
                      behind a record of their travels on the Kensington Runestone. No records of
                      any unusual astronomical phenomena. Large asteroid passed in 1362 AD.


      (40 years over ecliptic)

| | |
|---|---|
| 1400-1403 AD (5294-5297 AM) | In 1400 AD a meteorite crashed into an area close to Elbogen, Bohemia. It was metallic and still weighed 235 lbs.(2)  This may have been debris from the Spear of NIBIRU Apocalypse Comet. A large asteroid in the group passed over ecliptic in 1402 AD. |

(52 years under ecliptic completing 92 year orbit)

| | |
|---|---|
| 1452-1455 AD (5346-5349 AM) | Sultan Mehman II, known as Mohammed the Conqueror, conquered Constantinople, the Greek Christian capitol of the Eastern Orthodox Empire. The fall of this famous city was preceded by strange omens and unusual red lights over the Cathedral of St. Sophia. Something eclipsed the Sun and a terrible thunderstorm and torrential rain followed a strange mist. In 1454 a large asteroid passed over ecliptic. |

(40 years over ecliptic)

| | |
|---|---|
| 1492-1495 AD (5386-5388 AM) | This coincides with the passing of the Reuben Comet Group through the inner system. In this year Columbus and his crew were halfway across the Atlantic Ocean when a meteor burst and streaked across the sky. In 1494 a giant asteroid passed over the ecliptic, this being 552 years (Phoenix Cycle) before 2046 AD. Columbus discovered American islands. |

(52 years under ecliptic completing 92 year orbit)

| | |
|---|---|
| 1544-1547 AD (5438-5441 AM) | In 1546 AD a large asteroid unseen from Earth passed across the ecliptic beginning a 500-year countdown to impact on North America. Otherwise no records from this period. |

(40 years over ecliptic)

| | |
|---|---|
| 1584-1587 AD (5478-5481 AM) | If related it is not known, but in 1585 AD the Roanoke Colony on a North Carolina island (sponsored by Sir Walter Raleigh), *vanished*. Virginia Dare, the first person born in America from Europe (on record) was among the missing. Known as the Lost Colony, the mass disappearance baffled the local Indians and the Europeans who arrived later to investigate the site. Meals and all belongings were still left where they were set. In 1586 a volcano at Java, Indonesia killed 10,000 people and the huge asteroid of this group passed unseen over the ecliptic, this being 520 years (52 x 10: 52 being key prophetic number for America) to 2106 AD (6000 Am). |

(52 years under ecliptic completing 92 year orbit)

| | |
|---|---|
| 1636-1639 AD (5530-5533 AM) | Unseen from Earth, a huge asteroid passed over the ecliptic. No records for this period. |

(40 years over ecliptic)

| | |
|---|---|
| 1676-1679 AD<br>(5570-5573 AM) | In 1676 the New England Indian War ended after several bloody exchanges. In 1678 AD a large asteroid passed over the ecliptic unseen from Earth. In this year of 1678 appeared a crop circle formation, the first recorded and documented. A woodcut from this year depicts a strange figure in a field making the circle formation, which has come to be known as the Mowing Devil. |

(52 years under ecliptic completing 92 year orbit)

| | |
|---|---|
| 1728-1731 AD<br>(5622-5625 AM) | This date coincides with the appearance of the *2046 AD NIBIRU Comet Orbit*. From 1730-1735 earthquake activity was unceasing in the Canary Islands. A quake at Hokkaido, Japan killed 137,000 people. In 1730 a huge asteroid passed over the ecliptic unseen from Earth. |

(40 years over ecliptic)

| | |
|---|---|
| 1768-1771 AD<br>(5662-5665 AM) | In April, 1767 AD, over Germany (3) appeared an oblong, sulfurous cloud in the sky. In 1768 AD a meteor burst through the atmosphere over North America illuminating the night sky for 20 seconds, which the Shawnee called Panther Passing Across, which is literally *Tecumseh*, the name of a child born at that instant, who grew up to be the famous Indian leader of that name. He led several native American cultures in a war against European settlers.(4)  In 1770 the Vials of Phoenix Comet group began passing through the inner system and the huge asteroid of the 2046 AD Asteroid Impact group passed unseen over the ecliptic. Comet Lexell passed within 1.39 million miles from Earth (5), the closest pass of a comet in known historical times (the Sun is 93 million miles away). Comet Lexell approached Jupiter and fragmented.(6) |

(52 years under ecliptic completing 92 year orbit)

| | |
|---|---|
| 1820-1823 AD<br>(5714-5717 AM) | In 1820 a large object was seen traversing across the Sun. This transit is recorded on April 27th by astronomer Steinheibel of Vienna.(7)  In October of 1821 it rained a tremendous volume of an alien silky substance over France, a material *not* identified as gossamer.(8)  On November 27th, a bright luminous object traveled across the sky over Italy, an earthquake *following it* along the path it moved.(9)  In 1822 AD the huge asteroid passed over the ecliptic unseen from Earth and a rain of bituminous matter fell upon Norway.(10)  An earthquake afflicted Java killing 4000 people with eruptions lasting five hours.(11) |

(40 years over ecliptic)

| | |
|---|---|
| 1860-1863 AD (5754-5757 AM) | On January 1 an organic, combustible material fell upon Hessle, Sweden. (12) On April 11th astronomer Lias from Pernambuco, at noon on a cloudless sunny day witnessed an unexplainable darkening of the sky and the Sun was eclipsed. At this time Venus was at low visibility, but was observed to shine brightly. Most unusual is the fact that around the Sun appeared a corona. (13) In June a violent storm occurred and angular black pebbles rained upon Wolverhampton, England so abundantly that they had to be shoveled away. (14) In July at Dhurmsalla, India a meteorite fell that was *coated in ice*, and a few months later a reddish substance fell from the sky there. Astronomers noted a dark spot passing over the Sun and then an earthquake and a *yellow darkness* was followed by luminous objects in the skies.(15) Many meteorites fallen in India looked a lot like cannon balls of different sizes.(16) On December 28th, early in the morning, it rained a red liquid copiously for two hours over the northwestern part of Siena. Another red shower occurred at 11 AM, followed by a red rain three days later on the 31st.(17) In 1861 a carbonaceous matter fell at Cranbourne, Australia.(18) In 1862 blackish liquid rains fell at several places.(19) A comet appeared in 1862 with a very nebulous tail and black spots were recorded to have passed across the Sun. Among these may have been the large asteroid crossing the ecliptic in this year on its journey toward a 2046 AD impact. In 1863 AD at Manila, an earthquake was accompanied by the appearance in the sky of strange luminosities.(20) Over the Mediterranean in May fell a reddish rain and in October a black rain fell at Slains, Scotland.(21) |

(52 years under ecliptic completing 92 year orbit)

| | |
|---|---|
| 1912-1915 AD (5806-6809 AM) | This train may have spread somewhat, for in the year preceding, in 1911 AD on January 20th, a black rain fell on Switzerland and very conspicuous bright spots were seen on Mars (volcanoes?) (22) In 1912 on January 27th an intensely black object was studied that appeared over the Moon, which was estimated at 250 miles long and 50 miles wide. Clouds cut off further observation.(23) On March 6, 1912 a large fiery "machine" traveled across the English town of Warmley, and it traveled at a high velocity from the direction of Bath and was moving toward Gloucester.(24) Evidence that this may have been an Anunnaki vessel is perceived in that the object never hit the ground, nor was found. Further, on April 8th, a month later, a Charles Tilden Smith in Wiltshire, England near the *Stonehenge* site saw two large dark triangular shadows of objects concealed within rapidly moving clouds. As the clouds came and went in a continuous stream, the two shadows remained fixed for more than half an hour.(25) The Alaskan Mount Katmai erupted in 1912 destroying 50 square miles of forest. In 1913 on February 9th a tremendous procession of objects was seen passing 30 miles over the surface of the Earth from Canada, the United States, the Atlantic, and from Bermuda, of a luminous body with a long tail that grew rapidly larger and larger. It was in fragments, and they ". . .moved with a peculiar, majestic, dignified deliberation. It disappeared in the distance, and another group emerged from its place of origin. Onward they moved, at the same deliberate pace, in twos, or threes or fours."(26) Charles Fort reports that there were, in all, 30-32 bodies in the train and spectators related that the event appeared like a celestial fleet of vessels, like battleships attended with cruisers and destroyers. On April 8th, 1913, exactly a year after the April 8th sighting of two strange vessels hiding in clouds over Wiltshire in 1912, a shadow of an unknown object passed over the Moon as seen from Forth Worth, Texas.(27) In 1914 the huge asteroid that will impact North America passed over the ecliptic unseen from Earth. Japanese volcano Sakurajima erupted and typhus killed millions in Russia and Europe. |

(40 years over the ecliptic)

1952-1955 AD
(5846-5849 AM)
In July UFOs were visibly seen by the thousands over Washington, DC and detected on radar. Air Defense was initiated and the event was broadcast around America in headlines.(28) In 1953 astronomers discover large objects in space moving toward Earth. Believing them to be asteroids, they are reported to government officials. But the objects at a distance from the planet turned and adopted a geosynchronous orbit around Earth's equator. A dialogue between the occupants (Anunnaki) and government elitists was kept secret and a *trade-alliance* was initiated.(29) By June 1953 over 250 tornadoes had afflicted North America. In 1954 the large asteroid passed over the ecliptic unseen from Earth. This same year coincides with the return of the Spear of NIBIRU Apocalypse Comet into the inner system and is the year Elizabeth Anne Hodges was hit in the hip by a meteorite in her home.

(52 years under ecliptic completing 92 year orbit)

2004-2007 AD
(5898-5901 AM)
The asteroid Apophis was discovered in this year. The world's most powerful quake in 40 years happened underneath the Indian Ocean and caused a tidal wave that killed 200,000 people along the coasts of Southern Asia and Indonesia. The quake was near Sumatra and measured 9.2 in magnitude. The eventual death toll was estimated at 300,000. Twelve nations were afflicted. In 2005 an earthquake killed 80,000 people in India and Pakistan and hurricanes Katrina and Rita devastated American coasts with a death toll in New Orleans that has been suppressed. The published death toll does not at all correspond with the total amount of people still missing from Louisiana. Hurricane Katrina was accompanied with 33 tornadoes. In 2006 a meteorite visible for hundreds of miles entered the atmosphere and collided into a mountain in Reisadalan, Norway with a force equal to the Hiroshima bomb. Between July 2nd and 3rd, a large asteroid half a mile wide *ascended from below the ecliptic* and passed between the space between the Earth and Moon and was unexpectedly seen by astronomers who instantly reported it. It was traveling approximately 40,000 mph. Initial reports were aired on ABC during Good Morning America but within 15 minutes all subsequent reports focused only on Apophis, which was discovered in 2004. This information has been censored ever since. Why Apophis was heavily focused upon instead of this new, unknown asteroid hints that its existence was not meant to be publicly known. This was the huge asteroid that will *impact* in 2046 AD. It will orbit the Sun for 40 years and return to collide into North America. An earthquake at Jakarta, Indonesia killed 5000 people and triggered the volcanic eruption of Mount Merapi. 2006 AD is the 104th year (52 + 52) of Giza Course Countdown, which began in 1902 AD. Fifty-two years before the 2046 AD asteroid impact was the year 1994, when comet Shoemaker-Levy 9 collided into Jupiter. In 2007 another asteroid passed close to Earth and was widely reported to have been about 500 feet across. This news quickly died out.

(40 years over ecliptic)

| | |
|---|---|
| 2044-2047 AD<br>(5938-5941 AM) | In 2046 the giant asteroid will collide into the United States of America in the nations' 270th year from 1776 AD, exactly 60 years prior to the Armageddon War in 2106 AD (6000 AM). |

## Giza Impact Comet

The final comet to be detailed is a long-period object having broken away from NIBIRU in 1260 AD, entering its own elongated orbit around the Sun. It is called the Giza Impact Orbit because of the years this comet appeared and was seen, along with its relation to the mathematics incorporated into the dimensions of the Great Pyramid. Its total orbital length was 440 years, but it never made a second orbit.

| | |
|---|---|
| 1304 AD<br>(5198 AM) | This was during the 16 years of the Seven Comets over Europe, a period of earthquakes, plague fogs, falling stars and comets from Europe to China. |

(250 years over the ecliptic)

| | |
|---|---|
| 1554 AD<br>(5448 AM) | A comet appeared and was recorded by European astronomers.(30) This Anno Domini year of 1554 is the *Annus Mundi* year 5448, this number paralleling the vertical height of the Great Pyramid in pyramid inches (see *Chronotecture*). The Great Pyramid was the pre-flood world's 0 degree *prime meridian* and the structure contains within it an elaborate geometrical timeline with its passage and chamber dimensions that details historical periods both forward and backward in time. In this amazing year of 1554 AD, Mercator published his Great Map of Europe, correcting many of the cartographic mistakes of his predecessors.(31) Gerardus Mercator's research was not merely concerning the size and distribution of Earth's land-masses, but also its *history*. As the Great Pyramid served also an astronomical function, so too did Mercator embark upon a project in this year of unprecedented importance – he compiled a history of the world using the astronomical records of civilizations spanning back into distant antiquity.(32) This year 1554 is the 3000th year since the Flood in 2239 BC (1656 AM) and exactly one Phoenix Cycle of 552 years before Armageddon in 2106 AD (6000 AM). As the 5448 P" height of the Great Pyramid is 552 P" under 6000, so too does the deepest area of the Great Pyramid exhibit this 552-year correlation in the 552 P" width of the Subterranean Chamber. |

(190 years under the ecliptic, completing its 440 year orbit)

| 1744 AD (5638 AM) | The Great Daytime Comet appears, a horrifying spectacle in the night skies that exhibited *seven tails*. In England there was mass apocalyptic fear that the world was going to end.(33) Astronomers named it Comet De Cheseaux, but this comet would be renamed Shoemaker-Levy 9 in the future, with astronomers unaware that this was the same comet. This Anno Domini year was 1000 + 744, the sum of 744 being represented in the dimensions of the Great Pyramid's apex which, once its Cornerstone is in place, will be a total height of 5814 pyramid inches. This is 186 P" short of 6000, and as the monument has four faces, 186 x *4* is 744. |
|---|---|

(250 years over ecliptic)

| 1994 AD (5888 AM) | On February 1st a 32 ft. wide meteorite entered the atmosphere and exploded over the Pacific with a force of ten times that of the Hiroshima bomb. It had come undetected and possibly a part of the train of the Giza Impact Comet. Astronomers call the long-period comet Shoemaker-Levy 9, not realizing that it is the same as comet De Cheseaux. Shoemaker-Levy 9 approached Jupiter and fragmented into 21 pieces before colliding into Jupiter in a series of planet-wide explosions, caught on satellite film from a probe that took the pictures at a much different angle than from the Earth. A flux tube was clearly seen for the first time. In June, an earthquake in Bolivia of 8.2 magnitude shook much of South America and was felt in tall buildings as far north as Toronto, Canada. Los Angeles suffered extensive damage from a major quake and Mount Merapi in Jakarta, Indonesia erupted with a gas cloud of intense heat that killed 60 people. |
|---|---|

The only striking correlation this 1994-year has with the Great Pyramid is the number 52, which counts down in years to 2046 AD, when another asteroid and comet impacts *our world* instead of Jupiter.

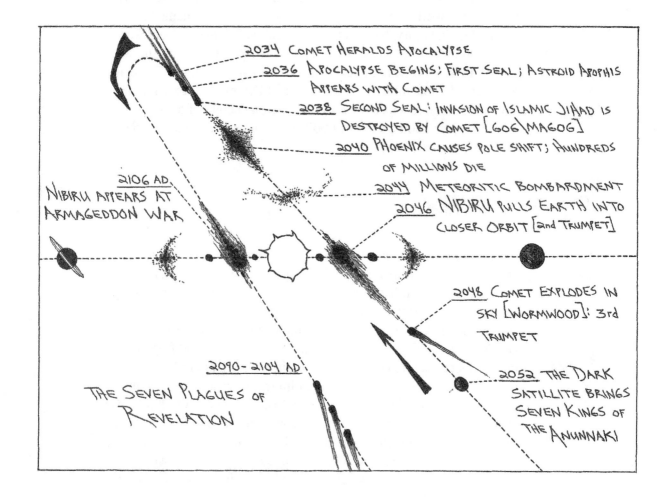

2034 Comet Heralds Apocalypse

2036 Apocalypse Begins; First Seal; Astroid Apophis Appears with Comet

2038 Second Seal: Invasion of Islamic Jihad is Destroyed by Comet [Gog\Magog]

2040 Phoenix causes pole shift; Hundreds of Millions die

2044 Meteoritic Bombardment

2046 Nibiru pulls Earth into closer Orbit [2nd Trumpet]

2048 Comet explodes in sky [Wormwood]: 3rd Trumpet

2052 The Dark Satillite brings Seven Kings of the Anunnaki

2106 AD
Nibiru appears at Armageddon War

2090-2104 AD
The Seven Plagues of Revelation

*Archive 16*

# Shards of NIBIRU

"Today astronomers specialize in branches of astronomy related
to space technology and have very little concern for historical
astronomy. . ."

—Dr. Arthur J. Brandonberger, Laval University, Quebec (1)

The pioneer of contemporary historical astronomy was Charles Fort, a researcher who diligently pored through the meticulous and vast volumes of scientists from around the world and correlated their findings in his books. These works have been the bane to scientists who refuse to accept Fort's life work. He is today virtually forgotten. To those remembering his works, he is regarded as an eccentric not to be taken seriously. And this, unfortunately, is the fate of Zechariah Sitchin, who opened our eyes to the on-goings and impending return of the Anunnaki. He is heeded by the masses that have read or listened to his theories while also kept at bay by his academic colleagues. These men and those that carry the torch afterward are regarded as imposters who trespass on the sacred terrain of science, charlatans seeking fanfare rather than the dissemination of truth. In 1932 Lewis Spence accurately commented on this situation (although referring to charlatans rather than serious researchers like Fort and Sitchin), writing:

"For every falsely inspired charlatan we have a hundred scientific
dullards, who, acting on the assumptions and acceptances of stupid
convention, actually believe that the mere collection of data suffices for
the foundation of a theory."(2)

Charles Fort and Zechariah Sitchin both agreed on one thing, and from entirely variant positions: a dark unknown world was periodically visiting the inner system, sometimes casting a shadow on our own planet, sometimes only on our Moon.(3) While Sitchin searched the annals of ancient Sumer, the Near East and the records of civilizations from antiquity to support his theory for NIBIRU, Fort diligently demonstrated from the pages of modern periodicals, newspapers, scientific reports and astronomical studies that strange objects were regularly entering the atmosphere, crashing in both remote and urban regions and that the inner system was known by astronomers to be populated by incredibly large and formerly unknown objects, as well as apparent constructions that had their own locomotion. . . Anunnaki vessels perhaps, although he stopped just short of stating this.

This author believes that he himself has provided the necessary *bridge* between the archaic evidence of Sitchin and the modern findings of Fort and others. It has become imperative that we have a working model of these theories in order to convince the critic of their probability. So much historical data concerning astronomical events is listed within this one book that it became apparent merely by a chronological organization of these accounts that Sitchin's Anunnaki Homeworld, NIBIRU, not only

141

exists, but has been seen throughout history and has seeded our solar system with comet groups and trains of fragments so many times and with such periodicity that these orbital periods were easily ascertained.

But this does not infer the perfection of a theory. There are some astronomical reports and earthly occurrences that have not appeared in these orbital chronologies because they did not fit. This does not make them less important, but demonstrates that not all objects in the inner system are attached to a major orbiting group. The solar system is now so full of junk that we cannot possibly calculate all of the things strewn throughout. It is these objects that frequently blot out more distant objects viewed through telescopes by astronomers, a phenomenon frequently reported but still unexplained.(4) Most astronomers neglect to report them for fear of ridicule, with the result for those who do report them being a condescending explanation concerning a flock of birds, satellite, weather balloon, swamp gas or a cloud, among other things.

The following is an abbreviated account of years not listed previously in this book and the unusual events that transpired.

| | |
|---|---|
| 1665 AD | In March a blue silky substance fell from the sky in great quantity near Naumberg.(5) A terrible plague stuck London, just as predicted by William Lilly. Official death toll in London alone was 68,596 people, but other estimates are closer to 100,000.(6) |
| 1669 AD | From March 8-11th a series of tremors and quakes coincided with the eruption from Mount Etna in Sicily, killing 100,000 people.(7) March 17th, on the town of Chatillon-sur-Seine, a reddish liquid rained that was thick and putrid.(8) |
| 1752 AD | On April 15th over Slavenge, Norway, was seen in the heavens a strange star of *octagonal* shape and balls of fire fell from a streak in the sky seen from Augermannland.(9) |
| 1755 AD | On June 7th a quake in northern Persia killed 40,000 people. On October 15th over Lisbon, Portugal were seen numerous meteorites (10) and sixteen days later on November 1st a transcontinental earthquake shook the lands and seas from the Baltic to the West Indies, and Canada to Egypt, and Algiers, Morocco, throughout northern Africa up to Germany, Holland, Spain and France. About 10,000 people were swallowed up by the earth and the world famous cosmopolitan city of Lisbon, Portugal, was covered in a great blackness (11) when the quake began ripping apart the city. About 60,000 people had taken refuge upon an expansive granite quay, which broke and plunged 600 feet deep, taking all the people with it as well as destroying entire ships full of cargo. Not a trace of these unfortunate people or vessels was ever found, nothing floated to the surface to betray their fate.(12) By 2 PM in England it was noted that the ocean had become restless with unusually high waves.(13) |

| | |
|---|---|
| 1783 AD | An earthquake at Calabria, Italy killed 30,000 people. The Icelandic volcano Laki erupted, as did the Japanese volcano Asama, and together the two, through dense atmospheric ejecta, initiated three successive extremely cold winters. The astronomer Herschel noted explosions on the Moon, indicating volcanic activity.(14) |
| 1833 AD | The Tuanaki islands in the Pacific vanished with all of its inhabitants (some reports put the date at 1844). Search vessels found no trace of the island or survivors.(15) A brilliant meteor show was seen over the United States in November. Many meteorites seemingly struck the ground but all that was found in the areas of descent was a strange, viscous gelatinous substance.(16) During the shower an unusual hook-like object traveling across the sky was seen over Ohio, and over Niagara Falls was seen a shining table-shaped object traveling with the meteor train, which appeared to stop for a while before moving on.(17) |
| 1839 AD | A massive continental earthquake spanning from Turkey into Russia occurred. Astronomer De Cuppis reported in October that he had discovered a massive object moving through the inner system.(18) Earlier on June 18th a profound darkness fell over Brussels and there fell flat pieces of ice, an inch long.(19) |
| 1852 AD | A triangular cloud on December 17 appeared during a storm. Within it was a red nucleus clearly visible, about half the diameter of the Moon and with a long tail. It was visible for thirteen minutes before the red nucleus exploded.(20) |
| 1866 AD | Four times in this year a black inky rain fell on Scotland and on June 30th a peculiar storm occurred over England. The people noted that the sky remained partially clear in areas, but rain and hail continued to fall. (21) Volcanic activity was recorded in the Aegean at Santorini. |
| 1878 AD | A total eclipse of the Sun occurred as the Moon transited and professors Watson and Rawlins of Wyoming witnessed two shining objects in the inner system near the Sun, a report verified by another witness, professor Swift of Denver, Colorado.(22) |

1882 AD        On January 22, at about 10:30 AM, a black darkness fell upon London,
               England and many other parts of Britain as reported by Major J. Herschel
               – so dark, in fact, that voices of people that could not be seen across
               the street were clearly heard. There was no trace of a fog.(23)  On July
               3rd people in Lebanon, Connecticut witnessed the appearance of two
               triangular shaped objects in the sky over the Moon. They moved toward
               each other and then vanished.(24)  From the diversity of other accounts
               it appears that these triangular shaped vessels have visited Earth several
               times recently and been seen. A gigantic cigar-shaped body was witnessed
               moving slowly across the skies through the telescope of the Royal
               Observatory of Greenwich, England on November 17th.(25)  It is the same
               form appearing as a shadow on satellite film from the 1989 Mars Observer
               probe right before the probe was destroyed. On December 21 astronomers
               noted a bright object near the Sun.(26)

1883 AD        Astronomers reported seeing a train of irregularly-shaped objects passing
               over the Sun.(27)  An earthquake on the island of Ishia off Naples was
               attended by a tumultuous sea.(28)  Quakes in Russia and Turkey dislodged
               great chunks of ice upon Mount Ararat. Turks explored the mountain and
               found the remains of a gigantic wooden ship fitting the dimensions of
               Noah's Ark, at over 12,000 ft. elevation.(29)  In February and March dense
               atmospheric dust was reported over Trinidad and Natal, South Africa.
               (30)  On February 27 an earthquake was felt in Connecticut as a luminous
               object was viewed in the sky.(31)  Meteorites fell during windstorms
               on May 18 at Hillsboro, Ill., on July 7th at Lachine, Quebec, and on
               September 24th in Sweden.(32)  On August 28th the volcano *Anak* (related
               in name to the biblical Anakim-Anunnaki?) on the Indonesian Island of
               Krakatoa exploded with a force of 200 twenty-megaton bombs, which
               excavated a hole at the bottom of the Pacific 1900 feet deep and ejected
               six-foot boulders as far as 25 miles into the atmosphere. The blast sent a
               tidal wave 100 feet high over the area of Java and Sumatra, destroying
               almost 300 towns and killing about 36,000 people.(33)  Volcanic ash
               was thrust into the atmosphere, and we are told that this dust-veil cooled
               the Earth, causing the next three severely cold winters and seven years
               of phenomenal sunsets and global hazes. However, much of this was
               caused not by volcanic ash, but by the earth being blanketed with *cosmic
               dust*, which was first reported in February and March, six months prior
               to the Krakatoa explosion. In November, loud noises were heard over
               Queensland, South Africa and there fell a pulpy substance like balls of
               dust.(34)  From November 10-12 was seen a strange comet with two
               unusual tails.(35)

1896

A submarine earthquake in the Tuscarora Deep, south of Japan, caused the sea to recede, leaving behind stranded ocean creatures. Multitudes of Japanese went out to gather the creatures and fish but they were quickly overtaken by a 90-foot tsunami returning to the coast at 450 miles per hour, killing up to 30,000 people. As the sea receded back from the land, people from even high villages were pulled out to sea to never be seen again. This was similar to the 365 AD event. No astronomical phenomena reported this year.

1907 AD

On July 2nd over the town of Burlington, Vermont, a terrific explosion was heard throughout the city and a ball of light was seen to fall from the sky, but disappeared. Some witnesses claimed to have seen some strange torpedo or cigar-shaped object in the sky.(36)  In October a black rain of a very disagreeable smell fell over Ireland.(37)

1908 AD

On March 27 a white substance like ashes rained upon Annoy, France, and on June 30th a bright, fiery object was seen by hundreds as it traversed the sky over Siberia where the object detonated over the heavily forested region of Tunguska. The blast knocked down trees and destroyed forest up to 300 miles away, creating a wasteland, even knocking people off their feet 100 miles away. Had the blast occurred over any city it would have ceased to exist. Iridium levels reveal that the object had approximately a 7 million-ton weight.(38)  For 10 days in early June, meteorologists recorded unusual intensity of the Northern Lights and other atmospheric disturbances. These signs were seen over Europe, Russia and Siberia, including the appearance of silvery mesospheric clouds, bright twilights, and extremely intense solar haloes increasing in intensity until the explosion. The explosion was seismically recorded in Irkutsh, Turkey, Tashkent, Tbilisi, and Jena and the pressure of the blast affected barometers in England. On July 2nd, exactly a year after the July 2nd explosion over Vermont in 1907, the Sun was shining and the sky was clear over Braemar when thunder was heard, and then pieces of ice fell from the sky.(39)

1989 AD

On March 23rd an asteroid over half a mile wide missed Earth by only 700,000 miles.(40)  It appeared undetected and had it appeared six hours later it would have probably killed 70% of all life on Earth.

One year before Sitchin's first book concerning NIBIRU was released in 1976, Dr. Bartholomew Nagy published his own controversial research in 1975 concerning the existence of microbiological fossils on and within *meteorites*.(41)  Like Sitchin's work, the evidence was ostracized from peer review and academic consideration. These two researchers essentially unveiled different aspects of the same truth. Sitchin reported on the existence of NIBIRU from old records while Nagy did so from an astronomical one, though he was unaware that the vast majority of carbonaceous materials falling from the sky were from NIBIRU. He was himself convinced these space fossils derived from Mars.

In 1950 a meteorite fell near Murray, Kentucky and two scientists of the Arizona State University independently examined the rock and detected the presence of all 18 of the known amino acids, as well as pyramidines from the nucleic acid of what were once *living cells*.(42)  In 1970 a meteorite fell to earth and was named the Murchison Meteorite. Studied by the Ames Research Center of NASA and the scientists of two universities, it was determined to have contained within it definite traces of amino acids.(43)  In 2001 scientists of the Ames Research Center released data on the discovery of carbon-rich meteorites found on Earth that were loaded with organic compounds, including amino acids and *sugars*.(44)

That NIBIRU was an inhabitable world supporting seas and vegetation is evidence by the study of those objects that fall to Earth. Our own world has been virtually enveloped in times past by viscous substances, silken webs (not from spiders), reddish and ink-black rains often smelling offensive, *real blood*, revealing that at some date prior to NIBIRU's virtual destruction, there had been a great war or disaster just prior to the surface of the Anunnaki Homeworld being quick-frozen. Bituminous matter has fallen to Earth, sometimes looking like coal and in other instances resembling a leafy, paper-like material. Decomposed vegetable matter that had been frozen for thousands of years, sulfurous meteorites, carbonaceous rocks, putrid resinous remains, downpours of pebbles, yellow and brown dusts and gelatinous pulp aerolites have all descended to our world from NIBIRU. . . sometimes in conjunction with actual vessels that periodically visit as well.

Which leads us to the end of our study.

An anciently destroyed world once inhabited by the Anunnaki and their own creations necessitates the existence of fauna and flora, if we are to accept the olden records that NIBIRU was a world like unto our own. Much of the decomposed and foul-smelling materials fallen to Earth throughout history may be the residual remains of such plants and animals that were encased in ice floating through space, but there is a phenomenon so totally ignored by the scientific community that this author has found it integral to this thesis on NIBIRU.

It is the origin of some insects, especially *arachnids*. Are they indigenous to this world?  In Sumerian and early American texts, the Anunnaki were represented as *scorpions* (arachnids), as seen in this author's work entitled *Descent of the Seven Kings*. In the Ten Plagues on Egypt, which has been demonstrated was linked to the appearance and tails of a comet, we find that an inordinate amount of insects plagued the Egyptians. Could they have come from space?  Insects used as a type of *judgment* upon mankind is a theme found throughout the biblical records. These little creatures are incredibly resilient, able under the most extreme conditions of cold and heat, humidity, submersion or aridity, to go into a stasis for years and years on end and when the weather-environmental conditions conducive to their survival and propagation return, they can come out of their bodily stasis and continue living as if nothing had ever transpired. A planet quick-frozen and fragmenting would have hundreds of billions of insects encased in its frozen surfaces that fracture off and become *comets*. Trapped in ice that enters the atmosphere and then through friction is changed to gas, many insects would survive the descent and then freefall upon the Earth, unlike mineral or metallic objects, which, through intense friction, burn up. Records of insect *rains* are by themselves unconvincing, however, the following various accounts, all collected by Charles Fort, have been selected to baffle even the critic.

| | |
|---|---|
| 1811 AD | Snow in Saxony was found to have a large amount of larvae from an unidentified insect species. |
| 1827 AD | During a snowstorm over Pakroff, Russia, fell immense numbers of black insects. |
| 1830 AD | As it snowed over Orenburg, Russia, a multitude of black insects said to look like gnats but with flea-like motions. fell from the sky. |
| 1837 AD | During a snowstorm over Bramford, Speke and Devonshire, England a large number of black worms, about three quarters of an inch in length, fell with the snow. |
| 1849 AD | On January 24th the weather was cold, no rain nor snow, but instead a rain of black larvae in enormous numbers over Lithuania. In the same year in August, after a large clap of thunder, a slab of ice 20 feet in circumference fell out of the sky at Balvullich, Ross-shire, England. It fell alone, without hail.(45) |
| 1850 AD | Large numbers of worms were found in a snowstorm near Sangerfield, New York. In Warsaw was found weird beetles that moved like caterpillars on the snow. |
| 1856 AD | Approximately 300,000 larvae fell with snow in Switzerland. |
| 1857 AD | A substance fell from the sky at Kaba, Hungary, which contained organic matter ". . .analogous to fossil-waxes."(46) |
| 1858 AD | In May, a large number of larvae beetles fell from the sky near Mortagne, France. The larvae were inanimate, as if with cold. |
| 1876 AD | Throughout Norway were discovered worms all over the snow and ground. This proved to be a great mystery, for the worms could not have come from the ground because it was *frozen solid*. |
| 1890 AD | In January a storm occurred over Switzerland, a tempest of *larvae*. They were black and yellow and incalculable. A great many birds were attracted to the area and fed upon the larvae. The yellow larvae were three times the size of the black ones. |
| 1891 AD | Worms fell with snow several times over Virginia, in the United States. |

1899 AD          Yellow and black worms were discovered on a glacier in Alaska, a region
                 where no plant or animal life survives.

1911 AD          On June 24th at Eton, Buckinghamshire, England masses of jelly fell from
                 the sky, the size of peas after a heavy rainfall. The substance was said to
                 contain eggs, and later small larvae emerged.

How interesting to find that the very celestial mechanisms that end life on Earth – comet impacts and detritus from the skies – also *brings life to earth*. Perhaps we would do well to reexamine much of the biblical records, like this statement God made to Job:

> "Has thou entered into the treasures of the snow?
>     Or hath thou seen the treasures of the hail,
> Which I have reserved against the *time of trouble*,
>     Against the day of *battle and war*?"

—Job 38:22

The disasters of the Book of Revelation and Old Testament prophecies will be delivered to our world frozen, ancient time capsules from the beginning that will initiate the judgments of the end. . . the Shards of NIBIRU.

# **Bibliography of Cited Works**

*The Science of God*: Gerald Schroeder (Free Press)

*The Herder Dictionary of Symbols*: English trans. Boris Matthews (Chiron)

*Symbols, Sex and the Stars*: Earnest Busenbark (Book Tree)

*From the Ashes of Angels*: *Forbidden Legacy of a Fallen Race*: Andrew Collins (Bear & Co.)

*Early Man and the Cosmos*: Evan Hadingham (Walker & Co.)

*The Iliad*: Homer (Barnes & Noble)

*Enuma Elish: Seven Tablets of Creation*: L.W. King 1902 Vol. I & II (Book Tree)

*The Book of Enoch*: trans. Richard Lawrence (Artisan)

*The Book of Enoch*: trans. R.H. Charles 1912 (Book Tree)

*The Book of Jasher*: (Artisan)

*The Book of Jasher*: trans. Albinus Alcuin 800 AD (Book Tree)

*Apocalypse of Baruch*: (Destiny)

*The Holy Quran*: (Presidency of Islamic Researches)

*Doctrine of Sin in the Babylonian Religion*: Julian Morganstern (Book Tree)

*Tracing Our Ancestors*: Frederick Haberman (Covenant Pub. London)

*Antiquities of the Jews*: Flavius Josephus trans. 1736 William Whiston (Henrickson Publishers)

*Dake Annotated Reference Bible*: Finnis Jennings Dake (Dake Pub.)

*Darwin's Mistake*: Hans Zillmer (Adventures Unlimited)

*Evolution Cruncher*: Vance Ferrell (Evolution Facts, INC)

*The Gods of Eden*: William Bramley (Avon)

*Gods of Eden*: Egypt's Lost Legacy and the Genesis of Civilization: Andrew Collins (Bear & Co.)

*The Cosmic Code*: Zechariah Sitchin (Avon)

*The Wars of Gods and Men*: Zechariah Sitchin (Avon)

*When Time Began*: Zechariah Sitchin (Avon)

*The Histories*: Herodotus: trans. Aubrey de Selincourt/revised John Marincola (Penguin)

*The Way of Hermes*: trans. Clement Salaman, Dorine Van Oyen, William D. Wharton and Jean Pierre Mahe (Inner Traditions)

*Poleshift*: John White (ARE Press)

*Lost Cities of China, Central Asia and India*: David H. Childress (Adventures Unlimited)

*Ancient Structures*: William Corliss (Sourcebook Project)

*The Great Pyramid: Its Divine Message*: D. Davidson and H. Aldersmith, 1924 (reprint Book Tree)

*The Destruction of Atlantis*: Frank Joseph (Bear & Co.)

*The Stones and the Scarlet Thread*: Bonnie Gaunt (Gaunt)

*Mythology*: Edith Hamilton (Little, Brown & Co.)

*The Greek Myths*: Robert Graves (Penguin)

*The Natural Genesis*: Vol. I 1883 Gerald Massey (Black Classic Press)

*Ancient Egypt Light of the World*: Vol. I/II Gerald Massey (Black Classic Press)

*The Light of Egypt*: The Science of the Soul and Stars: Vol. I/II Thomas Burgoyne 1889 (Book Tree)

*Necronomicon*: trans. 1977 (Avon)

*Temple of Wotan*: Ron McVan (14 Word Press)

*The Problem of Lemuria*: Lewis Spence 1932 (Book Tree)

*Voyages of the Pyramid Builders*: Robert Schoch and Robert McNally (Tarcher Putnam)

*Atlantis in America*: Lewis Spence 1925 (Book Tree)

*Far Out Adventures*: edited David H. Childress (Adventures Unlimited)

*Cataclysm!*: *Compelling Evidence of a Cosmic Catastrophe in 9500 BC*: D.S. Allen and J.B. Delair (Bear & Co.)

*Dark Star*: Andy Lloyd (Timeless Voyager Press)

*America and Great Britain: Our Identity Revealed*: William Dankenbring (Triumph Pub.)

*Three Books of Occult Philosophy*: Henry Cornelius Agrippa, annotated by Donald Tyson (Llewellyn)

*Predictions for a New Millennium*: Noel Tyl (Llewellyn)

*Ancient Civilizations*: (Greenhaven Press)

*Commentaries on the Occult Philosophy of Agrippa*: Willy Schrodter (Samuel Weiser)

*The Secret Symbols of the Dollar Bill*: David Ovason (Perennial)

*The Medieval Underworld*: Andrew McCall (Sutton)

*The Elegant Universe*: Brian Greene (Vintage Books)

*Tales of the Prophets*: *Great Books of the Islamic World*: Muhammed ibn abd Allah al-Kisai, trans. W.M. Thackston Jr. (Great Books of the Islamic World)

*Sages and Seers*: Manly P. Hall (Philosophical Research Society)

*The Secrets of Nostradamus*: David Ovason (Harper Collins)

*Prophets of Doom*: Damel Cohen (The Millbrook Press)

*Civilization or Barbarism*: Cheikh Anta Diop (Lawrence Hill Books)

*Round Towers of Atlantis*: Henry O'Brien 1834 (Adventures Unlimited)

*Atlantis: The Antediluvian World*: Ignatius Donnelly (reprint Book Tree)

*Atlantis in America*: *Navigators of the Ancient World*: Ivar Zapp and George Erikson (Adventures Unlimited)

*Secret Cities of Old South America*: Harold T. Wilkins 1952 (Adventures Unlimited)

*Great Disasters*: (Readers Digest Association, edited by Kaari Ward)

*The Stone Angle*: John Wun

*The End of Eden*: *The Comet that Changed Civilization*: Graham Phillips (Bear & Co.)

*Sacred Number and the Origins of Civilization*: Richard Heath (Inner Traditions)

*2007 World Almanac and Book of Facts* (World Almanac Pub.)

*The First Genesis*: *The Saga of Creation v. Evolution*: William F. Dankenbring (Triumph Pub.)

*Theogany*: Hesiod trans. Dorothea Wender (Penguin)

*The Histories*: Cornelius Tacitus trans. Alfred Church and William Brodribb (Penguin)

*Annals of Imperial Rome*: Tacitus trans. Michael Grant (Penguin)

*The Discoverers*: *A History of Man's Search to Know His World and Himself*: Daniel J. Boorstin (Vintage)

*The Later Roman Empire*: Ammianus Marcellinus: trans. Walter Hamilton (Penguin)

*Natural Hisjtory*: Pliny, trans. John F. Healy (Penguin)

*The Iliad*: Homer, trans. Ennis Rees (Barnes & Noble Classics)

*Behold a Pale Horse*: William Cooper (Light Technology Pub.)

*Early History of Rome*: Livy trans. Aubrey de Selincourt (Penguin)

*Secret Places of the Lion*: George Hunt Williamson (Destiny)

*Cosmic Codes*: Chuck Missler (Koinonia)

*Crop Circles: Signs of Contact*: Colin Andrews with Stephen Spignesi (New Page Books)

*Crop Circles: Gods and Their Secrets*: Robert Boerman (Frontier Pub.; Adventures Unlimited)

*The End of Days*: *Armageddon and Prophecies of the Return*: Zechariah Sitchin (William Morrow)

*The Great Pyramid*: *Its Secrets and Mysteries Revealed*: Piazzi Smyth 1880 (Bell Pub. New York. Reprint of 1978 AD Outlet Book Company, orig. published by W. Isbister, London, under title *Our Inheritance in the Great Pyramid*: 1880

*The Book of the Damned*: Charles Fort (Book Tree reprint)

*Our Haunted Planet*: John Keel (Glade Press 2002 reprint)

*In Search of Noah's Ark*: David Balsiger and Charles E. Sellier, Jr. (Sun Classic Books)

*Of Heaven and Earth*: edited Zechariah Sitchin (Book Tree)

*Mankind: Child of the Stars*: Max Flindt and Otto O. Binder 1974 (Ozark Mountain Publ.)

*The Dragon Legacy*: *The Secret History of an Ancient Bloodline*: Nicholas de Vere (Book Tree)

*Alexander the Great*: Robin Lane Fox (Penguin)

*The Merovingian Mythos*: And the Mystery of Rennes-le-Chateau: Tracy Twyman (Dragon Key Press)

*Sight Unseen*: *Science, UFO Invisibility and Transgenic Beings*: Budd Hopkins and Carol Rainey (Atria)

# Chapter-by-Chapter Notes

Archive 1.    *Existence and Return of NIBIRU*

1.    The End of Days 322
2.    Great Disasters 15
3.    Great Disasters 15
4.    Predictions for a New Millennium 63
5.    The Elegant Universe 339
6.    The Elegant Universe 79
7.    Evolution Cruncher 105
8.    Evolution Cruncher 99
9.    Evolution Cruncher 99
10.   San Diego Union Tribune July 20th, 2005
11.   Evolution Cruncher 107
12.   Cataclysm! 350
13.   Cataclysm! 204
14.   Book of the Damned 143
15.   Book of the Damned 145
16.   Dark Star 245
17.   2007 World Almanac
18.   Secret Places of the Lion 7
19.   Secret Places of the Lion 21
20.   Light of Egypt Vol. I 264
21.   Darwin's Mistake 160-161
22.   Darwin's Mistake 125-126
23.   Crop Circles, Gods and Their Secrets 119
24.   Light of Egypt Vol. I 136
25.   Light of Egypt Vol. I 138
26.   Light of Egypt Vol. I 139
27.   Light of Egypt Vol. I 264
28.   Temple of Wotan 58
29.   Atlantis in America 221
30.   Secret Teachings of All Ages 203
31.   Lost Cities of North and Central America 326
32.   Herder Dictionary of Symbols 187
33.   Necronomicon 16
34.   Doctrine of Sin in the Babylonian Religion 13
35.   Zoroastrianism 92-93
36.   Necronomicon 70
37.   Necronomicon 166

Archive 2.    *Orbital History of Planet NIBIRU*

1.    Atlantis in America 265
2.    Stones and the Scarlet Thread 51-52
3.    Stones and the Scarlet Thread 75, 45

4.    The End of Days 4
5.    The Great Pyramid: Smyth 370
6.    Book of Jasher 2:5-6, 19-20; Enoch 7:6-15, 8:1-9
7.    Ancient Egypt Light of the World Vol. II 567
8.    From the Ashes of Angels 183
9.    Book of Jasher 2:5-6
10.   Theogany: Hesiod 820-870
11.   Gods of Eden: Bramley 183
12.   Gods of Eden: Bramley 179-195
13.   Sages and Seers 10
14.   Gods of Eden: Bramley 181
15.   The Greek Myths 188
16.   The Medieval Underworld 159

Archive 3.      *Ancient Earth-Killer Comet Group*

1.    Tales of the Patriarchs 84
2.    The Naturel Genesis Vol. II 241
3.    Tacitus, Histories Book 5 p. 139
4.    End of Eden 157
5.    End of Eden 160
6.    2 Esdras 3:18-19
7.    Secret Places of the Lion 96
8.    America and Great Britain: Our Identity Revealed 334
9.    Civilization or Barbarism 82, citing trans. of Griffith from 1890
10.   Secret Cities of Old South America 46
11.   The Great Pyramid: Its Divine Message 348-349, Table XXVIII
12.   The Orion Prophecy 258
13.   Book of Jasher 90: 32-35
14.   The First Genesis 192
15.   Pliny, Natural History: Universe and the World 99 P. 22

Archive 4.      *King of Israel Great Orbit*

1.    1 Chronicles 21:1-5, 14
2.    1 Kings 2:10
3.    Tacitus, Annals I 26-28 p. 48
4.    Pliny, Natural History: Universe and the World 96 p. 21
5.    Voyages of the Pyramid Builders 213
6.    Voyages of the Pyramid Builders 214
7.    Voyages of the Pyramid Builders 214
8.    Stones and the Scarlet Thread 202
9.    2 Esdras 8:50
10.   2 Esdras 15:29-63

Archive 5.      *Orbital History of the Dark Satellite*

1.    Antediluvian World 190
2.    Pliny, Natural History: Mining and Minerals 27 p. 312
3.    Cosmic Codes: Messler 235
4.    Pliny, Natural History: Universe and the World 135-141

5.   Cosmic Codes: Messler 235
6.   2 Chronicles 32:31
7.   Ferrar Fenton trans. of Bible, 2 Chronicles 32:31
8.   Finnis Jennings Dake Bible note h on 2 Chronicles 32:31
9.   Secret Cities of Old South America 48
10.  Secret Cities of Old South America 39
11.  Herodotus, Histories Book II 142 P. 131
12.  Apocalypse of Baruch LXIII:8
13.  2 Kings 19
14.  Apocalypse of Baruch LXIII
15.  Pliny, Natural History: Universe and the World 89 p. 19
16.  Enoch 54:12
17.  Quran surah 18:94-100

Archive 6.     *Sodom-Trojan Apocalypse Comet Group*

1.   Josephus, Antiquities 1.6.4
2.   Ancient Structures 242
3.   Book of Jasher 11:10
4.   Ancient Structures 243; Lost Cities of China, Central Asia and India 244-245
5.   The Science of God 169; Poleshift 10
6.   Ancient Structures 244
7.   Evolution Cruncher 138
8.   Ancient Structures 243
9.   When Time Began: Age of the Ram: Sitchin
10.  Quran surah 69:6-7
11.  Quran surah 46:21, 24-25
12.  Tacitus, The Histories Book 5 p. 140
13.  End of Eden 140
14.  End of Eden 165-166, 167, note: Phillips believes comet appeared in 1486 BC but he is off by 47 years.
15.  The Destruction of Atlantis 169
16.  Homer, Iliad Book IV lines 446-449, 472-478
17.  Homer, Iliad Book I lines 184-188
18.  Homer, Iliad Book I 10-14, 50-60
19.  Homer, Iliad Book II lines 552-858 pg. 444 note 7, trans. Ennis Rees
20.  Homer, Iliad Book IV 82-91
21.  Homer, Iliad Book VIII 72-81, 146-151
22.  Homer, Iliad Book XI 54-58, Book XVI 524-528, Book XX: 61-68
23.  Mythology: Hamilton 183
24.  Livy, Early History of Rome 1.31
25.  Pliny, Natural History: Universe and the World 97 p. 22
26.  Pliny, Natural History: Universe and the World 147 p. 28
27.  Pliny, Natural History: Universe and the World 92 p. 20
28.  Pliny, Natural History: Universe and the World 96 p. 21, 99 p. 22
29.  Pliny, Natural History: Universe and the World 97 p. 22
30.  Pliny, Natural History: Universe and the World 92 p. 20
31.  Pliny, Natural History: Universe and the World 99 p. 22

Archive 7.        *Romanid Apocalypse Comet Group*

1.    Pliny, Natural History: Mining and Minerals 202 p. 364
2.    Pliny, Natural History: Universe and the World p. 22
3.    Pliny, Natural History: Universe and the World 200 p. 38
4.    Commentaries on the Occult Philosophy of Agrippa 67
5.    Voyages of the Pyramid Builders 216
6.    Voyages of the Pyramid Builders 216
7.    Pliny, Natural History: Universe and the World 100 p. 22-23
8.    Pliny, Natural History: Universe and the World 100 p. 23
9.    Pliny, Natural History: Universe and the World 99 p. 22
10.   Tacitus, Annals XII 41 p. 271
11.   Tacitus, Annals XII 57-58
12.   Tacitus, Annals XII 57-58 p. 278
13.   Tacitus, Annals XIII 20-23 p. 295
14.   Tacitus, Annals XIII 41 p. 304
15.   Tacitus, Annals XIII 10-12 p. 318
16.   Tacitus, Annals XIV 10-12 p. 318
17.   Tacitus, Annals XIV 21-22 p. 324
18.   Great Disasters 47
19.   Tacitus, Annals XV 20-22 p. 355
20.   Pliny, Natural History: Mining and Minerals 51 p. 328
21.   Tacitus, Annals XV 44 p. 367
22.   Antediluvian World 257
23.   Tacitus, Annals XVI 10-13 p. 387
24.   The Jewish War, Josephus 6.5.3
25.   Pliny, Natural History: Universe and the World 199 p. 37-38
26.   Tacitus, The Histories Book II p. 62-63
27.   The Jewish War: Josephus
28.   Ammianus Marcellinus Book 17:7:1-5
29.   Ammianus Marcellinus Book 19:12 p. 183
30.   Ammianus Marcellinus Book 22:13 p. 250
31.   Ammianus Marcellinus Book 25 p. 310
32.   Ammianus Marcellinus Book 23 p. 256, 291-292
33.   Ammianus Marcellinus Book 23 p. 255
34.   Great Disasters 55
35.   Ammianus Marcellinus Book 26 p. 333
36.   Three Books of Occult Philosophy 664 note 21
37.   Ammianus Marcellinus Book 30 p. 400

Archive 8.        *2046 AD NIBIRU Comet Orbit and the 2047 AD Reuben Comet Group*

1.    Round Towers of Atlantis 60
2.    Round Towers of Atlantis 60
3.    Round Towers of Atlantis 60
4.    Voyages of the Pyramid Builders 211
5.    Today, Tomorrow and the Great Beyond 156
6.    The Gods of Eden: Bramley 181-183
7.    Antediluvian World 33
8.    Book of the Damned 176

9.  Book of the Damned 176
10. Book of the Damned 176-177
11. Book of Fabulous Beasts 152
12. Sages and Seers 12
13. Sages and Seers 11
14. 2007 Almanac p. 669
15. The Discoverers 232
16. Book of the Damned 30
17. Book of the Damned 24-28
18. Book of the Damned 144
19. 2007 Almanac p. 270
20. Book of the Damned 175

Archive 9.  *Accurately Interpreting the Mayan Long-Count and its Relation to the Great Pyramid*

1.  Pages 60-61
2.  Our Haunted Planet 132
3.  The Destruction of Atlantis 193
4.  Pages 16-17
5.  Secret Cities of Old South America 408
6.  Ancient Mysteries 95
7.  The Lost Realms: Skywatchers in the Jungles: Sitchin
8.  Pliny, Natural History: Universe and the World 89 p. 19
9.  Round Towers of Atlantis 60
10. Book of the Damned 37
11. Ancient Structures 180
12. Ancient Structures 181
13. Ancient Structures 203
14. End of Mechanical Time: article by Jose and Lloydine Arguelles in Perceptions mag. July/Aug. 1995 p. 50
15. Psalm 144:12

Archive 10.  *Arrival of NIBIRU, Cataclysm and the Anunnaki Chronology of 2046 AD*

1.  Mythology, Hamilton 313
2.  Crop Circles, Gods and Their Secrets 119
3.  The Discoverers 66
4.  Cataclysm! 244
5.  Crop Circles, Gods and Their Secrets 124
6.  Evolution Cruncher 45
7.  Ezekiel 37:15-23
8.  2 Esdras 12:11
9.  The Vedas 92, Hymn 104
10. Symbols and Legends of Freemasonry 112
11. Temple of Wotan 58
12. Quran surah 78:17-23
13. Archeology and the Land of the Bible 133 Vol. II
14. The Natural Genesis Vol. I p. 298
15. Enuma Elish Vol. I line 28 First Tablet I. 132-136, Fourth Tablet

Archive 11.    *Condensed Calendar and the End of the Great Pyramid's*
               *Astronomical Chronology*

Archive 12.    *Stonehenge and the Chronometry of Original Time*

Archive 13.    *Stonehenge II, Crop Circles and the Anunnaki Necklace*

Archive 14.    *Noahic Warning Comet and the Spear of NIBIRU Apocalypse Comet*

10.     Medieval Underworld 159
11.     Sages and Seers 11
12.     Sages and Seers 9
13.     Cataclysm! 200
14.     Of Heaven and Earth: First chapter and editing by Zechariah Sitchin, data from pg. 139, Antonio Huneeus
15.     Destruction of Atlantis 173
16.     Antediluvian World 35
17.     Secret Symbols of the Dollar bill 126
18.     Of Heaven and Earth, First chapter and editing by Zechariah Sitchin data from pg. 139-140, Antonio Huneeus
19.     Of Heaven and Earth, p. 142 Huneeus
20.     Book of the Damned 37
21.     Book of the Damned 37
22.     Book of the Damned 37
23.     Book of the Damned 166
24.     Book of the Damned 153
25.     Book of the Damned 15-16
26.     Book of the Damned 74
27.     Book of the Damned 37
28.     Book of the Damned 56
29.     Book of the Damned 57
30.     Book of the Damned 207
31.     Book of the Damned 37
32.     Book of the Damned 30
33.     Book of the Damned 20
34.     Book of the Damned 178
35.     Book of the Damned 57
36.     Book of the Damned 51, 57
37.     Book of the Damned 49
38.     Book of the Damned 179
39.     Book of the Damned 20
40.     Book of the Damned 52
41.     Cataclysm! 280
42.     Book of the Damned 149
43.     Book of the Damned 30
44.     Book of the Damned 54, 22
45.     Book of the Damned 30, 48-49
46.     The Great Pyramid: Smyth 418
47.     Book of the Damned 135
48.     Book of the Damned 54, 162
49.     Book of the Damned 33
50.     Book of the Damned 80
51.     Book of the Damned 19
52.     Book of the Damned 19, 75
53.     Book of the Damned 57
54.     Book of the Damned 200
55.     Book of the Damned 210
56.     Book of the Damned 193

57.    Book of the Damned 170-171
58.    Book of the Damned 75
59.    Book of the Damned 55, 21
60.    Book of the Damned 208
61.    Book of the Damned 29
62.    Book of the Damned 51
63.    Cataclysm! 353
64.    Book of the Damned 139
65.    Book of the Damned 57
66.    Book of the Damned 51, 222
67.    Book of the Damned 211
68.    Book of the Damned 179
69.    Book of the Damned 155
70.    Book of the Damned 155
71.    Cataclysm! 352
72.    Book of the Damned 144
73.    Behold a Pale Horse 202
74.    End of Eden 196
75.    Our Haunted Planet 178
76.    Prophets of Doom 116
77.    The Cosmic Pulse of Life 18

Archive 15.    *Asteroid 2046 AD Impact Orbit Group and the Giza Impact Comet*

1.    Three Books of Occult Philosophy 238
2.    Cataclysm! 200
3.    Book of the Damned 177
4.    A Sorrow in Our Heart 39-41
5.    End of Eden 164
6.    Cataclysm! 200
7.    Book of the Damned 149
8.    Book of the Damned 44
9.    Book of the Damned 177
10.    Book of the Damned 56
11.    Antediluvian World 33
12.    Book of the Damned 57
13.    Book of the Damned 178-179
14.    Book of the Damned 129
15.    Book of the Damned 180
16.    Book of the Damned 190
17.    Book of the Damned 30-31
18.    Book of the Damned 57
19.    Book of the Damned 22-23
20.    Book of the Damned 179
21.    Book of the Damned 49, 23
22.    Book of the Damned 21, 144
23.    Book of the Damned 155
24.    Book of the Damned 191
25.    Book of the Damned 195

26. Book of the Damned 217
27. *Book of the Damned* 196
28. Kenny Young, MUFON
29. *Behold a Pale Horse* 202
30. Secrets of Nostradamus 157
31. The Discoverers 274
32. The Discoverers 274-275
33. End of Eden 163

Archive 16.  *Shards of NIBIRU*

1. In Search of Noah's Ark 46-47
2. Problem of Lemuria 181-182
3. Book of the Damned 167-168
4. Book of the Damned 138
5. Book of the Damned 44
6. Great Disasters 85
7. Great Disasters 91
8. Book of the Damned 31
9. Book of the Damned 177
10. Book of the Damned 177
11. Book of the Damned 182
12. Antediluvian World: (Book Tree) p. 40
13. Great Disasters 96
14. Book of the Damned 153
15. Riddle of the Pacific 291
16. Book of the Damned 37
17. Book of the Damned 210
18. Book of the Damned 147
19. Book of the Damned 139
20. Book of the Damned 75
21. Book of the Damned 23, 82
22. Book of the Damned 151
23. Book of the Damned 172
24. Book of the Damned 196
25. Book of the Damned 214
26. Book of the Damned 210
27. Cataclysm! 350
28. Secrets of Nostradamus 293
29. The First Genesis 262-263
30. Book of the Damned 13-14
31. Book of the Damned 179
32. Book of the Damned 130, 75
33. Civilization or Barbarism 69
34. Book of the Damned 54
35. Book of the Damned 211-212
36. Book of the Damned 213-214
37. Book of the Damned 21
38. Great Disasters 175
39. Book of the Damned 138

40.     Atlantis in America 102-103
41.     Far Out Adventures 459
42.     Mankind: Child of the Stars 26
43.     Mankind: Child of the Stars 25
44.     Sight Unseen 181
45.     Book of the Damned 147
46.     Book of the Damned 59

# About the Author

As of 2011 Jason M. Breshears has been in a south Texas prison for approximately 20 years, since he was 17 years old. He was given an agreed-to sentence that would require him to serve only seven and a half years in prison. In 1999 he was granted his parole release but Texas Parole Board adopted new retroactive policies that have since blocked his release from prison. Though a model prisoner and published author, Jason has been denied parole release five times and been made to serve over twelve years more than what his original plea bargain with the State mandated. His situation is not an anomaly in the draconian system of Texas politics. Until he is released he continues his research and writing, and at 38 years old has written the following works:

*Lost Scriptures of Giza: Enochian Mysteries of the World's Oldest Texts*
*When the Sun Darkens: Orbital History and 2040 AD Return of Planet Phoenix*
*Anunnaki Homeworld: Orbital History and 2046 AD Return of Planet NIBIRU*
*Descent of the Seven Kings: Anunnaki Chronology and 2052 AD Return of the Fallen Ones*
*Chronotecture: Lost Science of Prophetic Engineering*
*Chronicon: Timelines of the Ancient Future*
*King of the Giants: Mighty Hunter of World Mythology*
*The Book of Jason: Philosophical Musings of a Dark Prophet*
*2016 AD and Beyond: An Analysis of Nostradamus, Calendrical Isometrics and the Future*

For ordering information on titles so far published, see last page ad.

**ALSO FROM THE BOOK TREE**

# OF HEAVEN AND EARTH
*Essays Presented at the First Sitchin Studies Day*

## Introduced and Edited by Zecharia Sitchin

Six distinguished researchers, along with Sitchin, present evidence in support of his theories concerning the origins of mankind and the intervention of intelligence from beyond the earth in ancient times.

Presentations:

Zecharia Sitchin, *Are We Alone? The Enigma of Ancient Knowledge*

Father Charles Moore, *The Orthodox Connection*

M. J. Evans, Ph.D., *The Paradigm has Shifted – What's Next?*

Madaleine Briskin, Ph.D., *The 430,000± Years Pulsation of Earth: Is There a 10th Planet Connection?*

V. Susan Ferguson, *Inanna Returns*

Neil Freer, *From Godspell to God Games*

Jose Antonio Huneeus, *Exploring the Anunnaki-UFO Link*

They all agree that Sitchin's work is the early part of a new paradigm – one that is beginning to shake the very foundations of religion, archaeology and our society in general.

ISBN 1885395174 • 164 pages • 5.5 x 8.5 • paper • $14.95
1-800-700-TREE (8733)   www.thebooktree.com

# ALSO BY JASON BREASHEARS

***THE LOST SCRIPTURES OF GIZA: Enochian Origins of the World's Oldest Texts***, How and why was the Great Pyramid of Giza built? What was its purpose? What about the Sphinx? Are their clues today that will allow us to understand its meaning? Breshears takes us through these questions and provides interesting answers based on years of meticulous studies. His work brings us into various holy books, the study of ancient symbols and into the oldest known writings of ancient Egypt. After exploring every part of the pyramid both inside and out, the author reveals how it connects to history and how we can make future predictions based on the patterns he has deciphered. He advances the theory that all the major cataclysms of the past occurred at specifically predicable times, and when the next world-wide catastrophe will come. If his predictions are accurate – and history seems to bear this out – then we may need to prepare for what is coming. ISBN 978-1-585091119, paper, 8.25 x 11, 256 pgs, $24.95

***WHEN THE SUN DARKENS: Orbital History and 2040 AD Return of Planet Phoenix***. Numerous times throughout Earth's history there have been major cataclysmic events. These events have resulted in large-scale climactic changes, mythological stories of floods and visitations from the skies, and sometimes the complete extinction of life. The major planetary body that has caused much of this carnage has been referred to by many names. Jason Breshears has termed it Phoenix, based on his research into the distant past and what it was usually called by witnesses. By piecing together ancient documents from the most reputable sources available, we have, in this book, the most extensive and accurate rendering of the cycle of the Phoenix, including when it will come again. Some of us, according to the author, will live to see its return. Beyond the foundational scientific evidence, the author ties in various Bible prophecies that relate

directly to it. Many books exist on this subject, but few have broken new ground like this one, due to the extensive research involved. ISBN 978-1-58509-1171, paper, 6 x 9, 128 pgs, $14.95

Additional titles by this author may be forthcoming. Please direct inquiries or order above titles from:

**The Book Tree    1-800-700-TREE (8733)    www.thebooktree.com**

Printed in the USA
CPSIA information can be obtained
at www.ICGtesting.com
LVHW050321260923
759350LV00022B/298